HOMEBREW
BEYOND THE BASICS

HOMEBREW

BEYOND THE BASICS

REVISED & EXPANDED EDITION

➡ **All-Grain Brewing & Other Next Steps**

MIKE KARNOWSKI

STERLING EPICURE

New York

STERLING EPICURE
New York

An Imprint of Sterling Publishing Co., Inc.
1166 Avenue of the Americas

ISBN 978-1-4549-2810-2

Distributed in Canada by Sterling Publishing Co., Inc.
c/o Canadian Manda Group, 664 Annette Street
Toronto, Ontario M6S 2C8, Canada
Distributed in the United Kingdom by GMC Distribution Services
Castle Place, 166 High Street, Lewes, East Sussex BN7 1XU, England
Distributed in Australia by NewSouth Books
45 Beach Street, Coogee, NSW 2034, Australia

For information about custom editions, special sales, and premium and corporate purchases,
please contact Sterling Special Sales at 800-805-5489 or specialsales@sterlingpublishing.com.

Manufactured in China

2 4 6 8 10 9 7 5 3 1

sterlingpublishing.com

Cover design by Elizabeth Mihaltse Lindy

Additional photo credits: iStock: © Tim Awe: 115; © Peter Bocklandt: 66; © Cunfek: 80; © Brent Hofacker: 93;
© JHK2303: 36; © Karandaev: 107; © LICreate: 192; © Briana May: 88; © njpPhoto: 190; © Nordroden: 30;
© rasilja: 27; © Veronika Roosimaa: 79; © sadetgr: 141; © Whitestorm: x; © zmurciuk_k: 119;
Shutterstock.com: 122; Benoit Daoust: 76, 91; Roger Siljander: 138

To my sweet wife, Gabe,
who supported my brewing
obsession from the start

CONTENTS

INTRODUCTION

It's hard to believe that so much could change in homebrewing in the four years since I wrote the first edition of this book. New styles such as the New England IPA have emerged, new hop varieties have exploded in popularity, and new advances in sour-beer production have changed the landscape. I'm pleased to address these new advances in ingredients and techniques in this revised and expanded edition.

Most homebrewers start out the same way: They begin with a simple malt-extract kit, progress to steeping crushed specialty grains, and eventually try partial mashing combined with extract. These methods can result in perfectly fine beer, and many brewers short on time or space may never go any further—which makes sense, in a way. If you can make a great beer through extract-based brewing, why make the jump to all-grain?

> **Welcome to all-grain brewing, a method of beer making that puts you in complete control of the creative process.**

Well, extract-based brewing is a bit like making macaroni and cheese from a boxed mix: You can play around with the recipe, add plenty of butter and fresh cheese, and end up with a delicious dish that nobody would guess came from a box. But eventually you'll start to feel that your creativity is being stifled by that premixed powder. You'll wonder, *What if I could choose whatever ingredients I want and cook them whatever way I want?* In short, you'll eventually want total control over your finished product.

Welcome to all-grain brewing, a method of beer making that puts you in complete control of the creative process. From the selection of grain and mash temperature to the mineral content of the water you're using (as well as many other variables), you're in charge when you do an all-grain brew. If you make beer using malt extracts, you leave these and other decisions to the extract manufacturers. But if you brew with grain, you have the capability of making a beer that's yours from step one.

This book shows you how to do that, starting with a look at brewing equipment and the steps you'll go through on brew day. The chapters that follow explore the many options you'll have as a brewer. We'll tackle malts and different ways to mash. (For most all-grain brewers, malts are just as exciting as hops.) We'll go in-depth on the topic of hops as well, with a look at how you can get the most from them. We'll also explore yeast varieties and the important practices surrounding them.

Sprinkled throughout the book are recipes for a variety of homebrews, including funky, sour, and wood-aged beers, which often are the hardest to find and the most expensive to buy. I'll walk you step by step through the brewing of each.

Finally, we'll look at a variety of nontraditional ingredients: fruits such as apricots, peaches, and cherries, as well as herbs and spices that can make your brews truly unique. Lovers of chocolate stouts, pumpkin beers, herb-infused pale ales, and anyone who brews off the beaten path will find much to savor.

Homebrew Beyond the Basics is geared toward the reader who has some knowledge of the beer-making process. It thoroughly explains key methods while touching lightly on more basic information. Truly basic info is easy to find elsewhere—online, for instance, or from other homebrewers.

This book is designed for and dedicated to brewers who want to make beer that's as good as or better than commercial beer. There's no real obstacle to that goal other than attention to detail and experience. If you follow the instructions in this book, there's no reason why you can't brew world-class beers.

ABOUT THIS BOOK

If you didn't just read the introduction, let me be clear:
This book assumes that you have some brewing knowledge.
It also makes some other assumptions:

➡ You have basic brewing equipment and know how to use it.
It's okay if you don't—maybe you've brewed only with friends.
But the equipment chapter is devoted to all-grain pieces only.
If you need any information on basic pieces (buckets, airlocks,
etc.), you can find it easily at your local homebrew shop or
online.

➡ All the information conveyed applies to the standard
homebrew batch size, 5 gallons. Chances are that will work for
you, but keep it in mind as you translate things to your own
system if you brew smaller or larger batches.

➡ The recipes in this book go hand-in-hand with the chapters
they're in by putting the information you just read into practice.
Sometimes that will mean trying out a technique, such as kettle
souring. Other times, a recipe will feature an ingredient, such as
home-toasted malt. Once in a while, a recipe will just give you a
platform to do your own thing (such as the Blonde Ale on page
180 designed as a base beer for fruit or herbs.)

➡ Speaking of the recipes, I did my best to make them easy
to understand and easy to modify. My efficiency is typically
between 70 and 75 percent, and all original gravity ranges
operate under that assumption. But if you plug it into your
own calculator and get a far different gravity or if you're
brewing a different batch size, just use the grain percentages.
The same goes for hops—I used alpha acid units (page 69)
rather than ounces to make things easy to customize based on
crop variations or hop substitutions.

GOING ALL-GRAIN

EQUIPMENT OVERVIEW

If you already brew with extract, there's not much holding you back from going all grain. Well, okay, it does cost money to buy some new equipment, but at least you already have all of the basic brewing equipment listed on the next page. Since 5-gallon batches are the standard size for all-grain brewing, just as they're for extract, most brewers jump straight from 5-gallon extract to 5-gallon all-grain. But if you're short on space or money, you may decide a smaller 3-gallon setup is right for you. If you go through beer quickly, you may want to move up to 10-gallon batch sizes when you invest in all-grain equipment. See the table below for the key differences in equipment, then keep reading for detailed information on the big pieces.

Setups for All-Grain Brewing

Setup	5-Gallon Extract Brewing Equipment	3-Gallon (Small Batch) All-Grain Brewing Equipment	5-Gallon All-Grain Brewing Equipment	10-Gallon (Double Batch) All-Grain Brewing Equipment
Boiling Kettle Size	3–5 gallons	4-gallon minimum	10-gallon minimum	15-gallon minimum (almost everyone uses a beer keg with the top cut off)
Second Kettle Size*	Not needed	3-gallon minimum	5-gallon minimum	10-gallon minimum
Mash Tun	N/A	5-gallon pot or cooler	5- to 10-gallon pot or cooler; the larger the mash tun, the more grain you can fit in, so if you plan on brewing strong beers, go big.	10-gallon pot or cooler
Heat Source	Stovetop	Stovetop	Either an outdoor propane burner (preferred) or a very strong stovetop	Outdoor propane burner
Wort Chiller Needed?	No	No	Yes	Yes

*Note: Most brewers use a second kettle to heat sparge water. Going this route will speed up your brew day, but if that's not possible, use several smaller pots or preheat the water in your kettle and store it in a cooler.

BASIC BREWING EQUIPMENT

$6\frac{1}{2}$-gallon bucket with lid (or 6-gallon glass carboy) to use as a fermentor

Airlock (with a bung if you're using a carboy)

Hydrometer (and/or refractometer)

Racking cane (or auto-siphon)

Siphon tubing

Thermometer

Cleaner (automatic dishwasher detergent, Powdered Brewery Wash (PBW™), or OxiClean™ is recommended)

Sanitizer (Star San™ or iodophor is recommended)

MASH AND LAUTER TUN

We'll go in depth on the mashing and sparging processes on page 44, but you probably already know that you're going to mix more than 10 pounds of crushed grain with quite a few gallons of hot water. The vessel you use for this is generally referred to as a tun. After letting it sit, you're going to run some additional hot water over it and rinse the grain. While some homebrewers and commercial brewers mash in one vessel and then transfer to a lauter tun, most use a tun fitted with a strainer or false bottom for both steps. The wort then goes into a separate kettle for boiling. On the homebrew level, you have two main choices for a tun: a cooler or a pot.

A **10-gallon cooler** is the most popular choice because it's fairly inexpensive, holds the temperature steady without much work on your part, and is easily fitted with a variety of strainer mechanisms. The downside of coolers is that you're pretty much limited to single-temperature mashes. (If it puts your mind at ease, the majority of commercial breweries have the same problem.)

A **5- to 10-gallon stainless steel pot**, on the other hand, allows you to apply direct heat to the mash—which lets you easily perform multistep mashes without any changes in water volume. (See page 50 for more on decoction mashes.) The downside of pots is that they lose heat more quickly than a cooler, and they're typically more expensive.

There are variations on these primary options—some of which utilize pumps—known as recirculation infusion mash systems (RIMS) or heat exchanger recirculating mash systems (HERMS). These are pretty complex and pricey. I've always been a fan of simple equipment and recommend a basic cooler or pot system. You can always incorporate pumps into your system at a later time if you get the itch. But you might never need them: I've used a simple pot and screen strainer for almost 20 years and been perfectly happy with them.

Valves

All standard coolers and many kettles come with a convenient hole for adding a valve. (For coolers, you just remove the standard plastic pouring spout.) If your kettle doesn't have a hole, first locate a valve and then drill a hole—not all valves are exactly the same diameter.

Most homebrew suppliers now sell valves with everything you need (all components and O-rings) in one package at a fair price. Shop around and find one that fits your needs. Stainless-steel parts are always the best option if you can afford them. If you decide to build the valve yourself, you can save a dollar or two, but you probably will have to order parts because most home improvement stores tend to stock inferior metal valves and not stainless steel.

Strainers

The right strainer for you will depend on your brewing setup and your budget. All strainers need to connect to your valve on the interior of your mash tun, so make sure to check that a particular strainer will connect to your valve before you buy it.

If you're mashing in a round picnic cooler and you have the money, go for a **false bottom**. If you plan to use a pot as a mash tun, I recommend a **screen strainer** as opposed to a false bottom. When you apply heat to the pot, you won't be able to stir any wort underneath a false bottom, and it might scorch. If you have a rectangular cooler, consider building a **copper manifold** because the far side of your mash would be pretty far from a small screen, and there's no readily available false bottom. That's the core of it, but feel free to read on and learn more about your options.

Let's start small. One of the simplest strainers is a rolled tube of stainless-steel screen (such as a window screen but made of stainless steel), crimped on one end, with the other end connected to the spigot exiting the mash tun. These screens are inexpensive and available at any homebrew supplier, or you can make your own. If you're just getting started or if you're short on cash after buying other equipment, this screen will get the job done

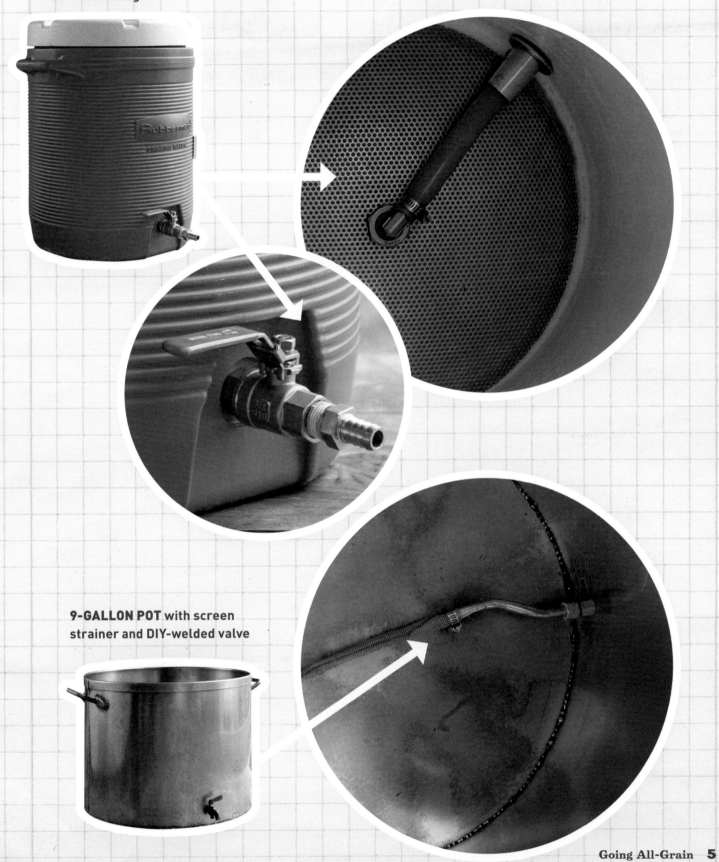

10-GALLON COOLER with false bottom
and store-bought stainless-steel valve

9-GALLON POT with screen
strainer and DIY-welded valve

for most mash tuns. The worst that will happen is a slightly lower efficiency (see below) than with other, larger filtering systems.

Another popular form of strainer, a false bottom, covers the entire bottom of a pot or cooler. Making your own false bottom is difficult, because it requires finding a sheet of stainless steel or food-grade plastic, cutting it to fit, and drilling hundreds of small holes in it. Luckily, manufactured false bottoms are available for most sizes of pots and coolers. Why spend the extra money? False bottoms are more durable than screens, which can wear down and even break due to excessive bending over time. False bottoms also drain uniformly over the entire surface, as opposed to a single strainer, which can leave sugars in the areas farthest away. But I wouldn't place too much importance on wort flow.

Widely used in the early days of homebrewing, copper manifolds now are found usually only in rectangular picnic coolers. In addition to fitting odd-size mash tuns, they're also cheap and easy to make. A few feet of copper tubing, a few copper elbows, and a hacksaw are all you need. The pattern of the manifold is up to you—just don't make it too complicated! The output pipe pushes through the hole where the spigot used to be, and a hacksaw will cut small slits in the copper pipe on the bottom side that sits on the floor of the cooler. Food-grade tubing is attached to the output pipe along with an inline valve for controlling the rate of flow. Everything you need is available at your local home improvement store.

MASH-TUN EFFICIENCY BATTLE

The three brewers at my brewery have different mash tuns for our homebrew systems, so we decided to put them side by side and see how they compared. We filled each tun with the same strike water and a 50-50 blend of wheat malt and German pilsner malt. After a 60-minute rest at 149°F, the mash was run off using fly sparging (page 54) until we collected 6½ gallons. These were the results:

Mash Tun #1 (9-gallon stainless steel pot with a thin, single-screen tube on the bottom): 1.043 specific gravity (SG)

Mash Tun #2 (9-gallon stainless-steel pot with a stainless-steel false bottom): 1.045 SG

Mash Tun #3 (10-gallon cooler with a homemade coil): 1.045 SG

To see what difference batch sparging (page 54) had on Mash Tun #1—the mash tun with the highest likelihood of missed sugars due to fly sparging—we repeated the experiment using batch sparging. Theoretically, this would help rinse the sugars on the grain hiding in the corners back into solution. The result was a slight increase to 1.044 SG.

What I take from this experiment is that most common mash-tun designs give you similar and acceptable results when it comes to efficiency. It's always nice when specialized and more expensive equipment turns out to have very little impact.

KETTLE

After you've drained the wort from the grain, you'll need a kettle large enough to boil the full amount of wort. Since that's close to 7 gallons for a 5-gallon batch, you'll want a pot that holds about 10 gallons. If your pot barely holds the amount of liquid needed, say 8 gallons, you'll undoubtedly have boilovers. If you get a boiling kettle that's cheaply made, you'll likely regret it later. In general, you want a stainless-steel pot with heavy-duty handles and a thick bottom (18-gauge) to help prevent burning.

It takes only a little more time to make 10 gallons of beer than it takes to make 5 gallons. If less work for larger batches sounds good to you, consider spending a little more on a pot now so that you have the option of brewing a 10-gallon batch down the road. If you want to be able to brew 10 or more gallons at a time, then you'll need a 14-gallon or larger pot.

Since stainless-steel pots in large sizes can get mighty pricey, many homebrewers make their own by cutting the top off of a stainless-steel keg to make a keggle. (It's known as a keggle because it's part keg and part kettle.) It's fairly easy to acquire a keg legally and cheaply; most breweries have a few sitting around that don't hold pressure, and they sell them at good prices to homebrewers. Please don't steal kegs! Cutting the top off a keg should fall to someone with the right tools for the job. Having cut a few kegs with grinders and jigsaws in my time, I can tell you it's worth the money to have a welder do it for you. Once the top is cut off, you have a 15 1/2-gallon stainless steel pot that's perfect for boiling 5- or 10-gallon batches.

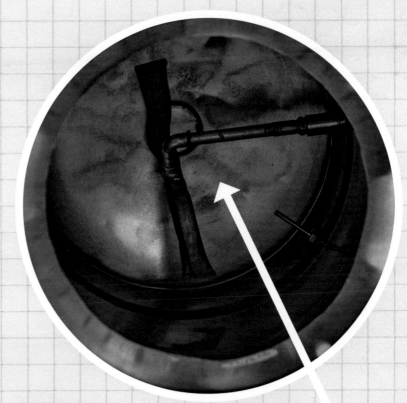

KEGGLE with thermometer, stainless-steel valve, and screen strainer

Burner

The only way to get a 10-gallon pot to a rolling boil on most stoves is to buy a wide pot and have it straddle two burners. You're rolling the dice on your stove strength when you order the kettle, but for some city-dwelling brewers, there's no other choice. If this is the boat you're in, look for a pot that's at least 16 inches wide (or wide enough to fit across most of two burners).

If you have a driveway or yard, it's probably time for you to take your brewing from the kitchen. Most stoves just aren't made to boil that much wort! Inexpensive propane burners are available at any home improvement store and work just fine for brewing. Try to avoid the "jet-burner" style that has just one main burner opening. A "banjo" style has many burners in a ring, which minimizes scorching and gives you more control over the boil. Also, if you can get a burner with taller legs, it will make life easier for you because you won't need to lift the kettle in order to siphon into your fermentor. A spare propane tank also is useful; running out of propane in the middle of a brew day is a huge pain.

WORT CHILLER

When you finish boiling a standard-size batch, you will have about 5½ gallons of very hot wort. Trying to chill that down in your bathtub isn't the way to go. You need something that will cool the wort quickly to yeast-pitching temperatures. You need a wort chiller.

A wort chiller can be as simple as a copper pipe bent into a coil and attached to a garden hose or as fancy as a commercially made plate chiller. Let's start with the most common style of chiller: An **immersion chiller** consists of a copper coil attached to a garden hose or other water source. The coil boils with the wort for the last 20 minutes in order to sanitize it. Then, when the heat turns off, you start the flow of cold water through the chiller—which moves through the inside of the coil and slowly cools the wort from the inside out. Brewers like immersion chillers because they're cheap (you can even make your own), sanitation isn't a worry because it's boiled with the wort, and it allows you to siphon off very clear wort that's mostly free of cold break. The only downfall of immersion chillers is that they tend to use a lot of

BANJO-STYLE BURNER
with propane tank

water. But if you run off the water you're using into your garden or washing machine, you can partially mitigate that.

Another popular type of chiller is a **counterflow chiller**. It takes hot wort from the kettle and runs cold water in the opposite direction, so your wort is cooled very quickly on the way to the fermentor. This can be a DIY project, but it's quite a bit more complicated than a simple immersion chiller. The benefit of a counterflow chiller is that it uses less water and is quicker than an immersion chiller. But it has two cons: It requires a meticulous cleaning and sanitizing regimen to prevent contamination, and the break material in the wort transfers into the fermentor instead of staying behind in the kettle.

Some commercially made counterflow chillers are based on the type of chiller that commercial breweries use. (In a brewery, the chiller is called a "heat exchanger" and looks like a large block made of many sheets of metal.) On a homebrew scale, these are referred to as **plate chillers**. As the hot wort passes through this maze

of sheets, cold water passes in the other direction on the other side of the sheets. Just as with regular counterflow chillers, cleaning and sanitizing are much more involved than for standard copper wort chillers.

REFRACTOMETER

You don't need this for all-grain brewing, but I can't believe I brewed for 20 years without a refractometer— and I can't imagine brewing without one now. With just a few drops of wort, I can check my sparge runnings to make sure I don't extract any tannins by going below 1.008. I also can quickly check the pre-boil gravity of my wort so I can stop sparging when I hit the gravity I want.

Note: A refractometer is great for checking original gravity (OG), but it's useless once the beer starts fermenting. The alcohol plays havoc with the readings, so after fermentation you need to switch back to a hydrometer.

A dual-display refractometer that shows both specific gravity (SG) and Brix scales is the most convenient, and you can buy one on eBay for a nominal amount. If you find one with only Brix/Plato units, you can use an online calculator to figure out the SG.

IMMERSION WORT CHILLER

CLEANING AND SANITIZING

CLEANING

While you can't say cleaning is more important than sanitizing, you can say it's impossible to sanitize something that isn't squeaky clean. Commercial brewers use very hot water (160–190°F) and strong caustic (alkaline) cleaners. Working with caustic cleaners is fairly dangerous, and you must use eye and skin protection, so they aren't recommended for homebrewing—especially in homes with small children.

A good alternative is an alkaline cleaning product that's a bit safer around the house. The easiest one to find is in the cleaning aisle of your local supermarket in the form of unscented liquid dishwasher detergent. Buy a cheap brand without any fragrances. You'll find it's an excellent and affordable cleanser. Just 2 to 3 tablespoons in 5 gallons of hot water works well for soaking plastic equipment. See the chart below for more on this and other cleaning-solution options.

I prefer letting chemicals and time do most of the work. I usually just let my equipment sit in a cleaning solution for a day or two and then rinse it well. Scrubbing with brushes or abrasive pads is a no-no; these cause micro scratches that bacteria can hide in, which makes sanitizing more difficult. Also, be careful about shocking glass carboys with hot water or going from hot to cold water quickly; they're very prone to breakage. Warm water in the 120°F range is fine and generally won't crack the glass. Rinse any cleaner multiple times before sanitizing, especially if you used bleach, which easily can add plastic (phenolic) aromas and tastes to your beer.

Note: Separate any moving plastic parts when cleaning and sanitizing. The plastic spigots on bottling buckets are notorious for harboring bacteria, as is the spring in a bottle filler. Look closely at your equipment. Is that a piece of hop flower stuck in your auto-siphon? Remember, if it's not clean, it can't be sanitized properly.

Cleaning-Solution Options

	Automatic Dishwasher Detergent (unscented)	Powdered Brewery Wash (PBW)	OxiClean	Bleach
Pros	Cheap, works well	Works well	Works well, especially good for soaking tough debris	Cheap
Cons	None, unless you accidentally buy one with fragrance	Expensive	More expensive than automatic dishwasher detergent	Can etch glass carboys at high concentrations; if not rinsed well it can leave major off-flavors
Dilution	2–3 tablespoons per 5 gallons of water	Follow directions on container.	1–2 tablespoons per 5 gallons of water	¼ cup per 5 gallons of water
Notes	Avoid versions with fragrance added.	Order in bulk online or through your local homebrew store for better pricing.	Generic versions work just as well.	In general, avoid it unless you're out of other cleaners and need one in a pinch.

SANITIZING

Commercial breweries traditionally use acid-based sanitizers. Just as with commercial cleansers, these are quite dangerous and will turn your skin white and blind you if you're not careful. These products aren't marketed to homebrewers for reasons of liability. However, a similar product marketed as Star San is highly recommended. Don't be concerned about the foaming quality of Star San (it creates large bubbles); a small amount of this sanitizer won't affect the taste of your beer. Star San is a no-rinse sanitizer, so you just need to shake out any residue, and you'll be good to go.

Iodophor is an iodine-based sanitizer that many homebrewers use with good results. Like Star San, it's no-rinse. The only downside is that iodophor tends to stain plastic parts red; this is only cosmetic and doesn't affect the beer.

Bleach isn't recommended as a sanitizer, because there's too great a chance of off-flavors if it's not rinsed extremely well. (If you're rinsing with tap water, then you just un-sanitized.) Stick with Star San or iodophor at the recommended dosage and contact time.

Boiling water or steam isn't practical for most brewing equipment, but stainless steel and some glass items (e.g., small jars) can be boiled for 20 minutes to sanitize. Make sure to cover the boiling pot and let the items cool to room temperature before handling. Direct flame also can be used to sanitize some smaller items, such as the lip of a bottle or the neck of a carboy. Be as careful with fire as you are with chemicals.

The most important thing to keep in mind is that bacteria and wild yeast travel on dust particles in the air. Covering the cooling wort in your kettle with a clean trash bag or using foil to cover the neck of a carboy will keep most anything from drifting into your wort and giving you problems. Once you get in the practice of proper cleaning and sanitizing, it'll become second nature, and you'll hardly have to think about it: clean (soak), rinse, sanitize, and keep covered.

BREW DAY

BREWING AN ALL-GRAIN BEER

A smooth brewing experience is all about preparation and practice. The more you brew, the more familiar you'll become with every stage of the process. For this example, we'll be brewing my house American Pale Ale. The basic process of mashing is covered within the directions, but you can find an in-depth explanation on page 45.

Note: Throughout the book, the recipes include targets. For more on original gravity (OG) and final gravity (FG), see page 13. For more on international bitterness units (IBUs), see page 69.

YOU NEED
basic brewing equipment (page 3)

9 gallons filtered brewing water (step 1)

10 pounds American 2-row

½ pound crystal malt (60°L)

½ pound Briess® Special Roast malt (50°L)

7.5 alpha acid units Columbus hops at 60 minutes (24.6 IBU)

7.5 alpha acid units Columbus hops at 30 minutes (18.9 IBU)

1 ounce each Chinook and Cascade (dry hops)

2 vials or packages California Ale WLP001/ American Ale WY1056 yeast (note on page 103)

Whirlfloc® tablet or Irish moss (optional)

TARGETS
Yield: 5 gallons

OG: 1.050–1.054

FG: 1.010–1.012

IBU: 43.5

STEP 1 ➡ PREPARE YOUR WATER.
This step can be done the day before you brew. For this recipe you need 9 gallons of brewing water: 4 gallons for mashing and 5 gallons for sparging. If you decide to brew another recipe for your first beer, a good rule of thumb is 1 gallon of water for every 3 pounds of grain for the mash, and the same amount of sparge water as your finished beer yield. So if you hope to have 5 gallons of beer at the end of the day, you would use 5 gallons of sparge water.

Brewing water needs to be free of any chlorine and other impurities. The easiest thing to do when starting out is to buy purified (ideally reverse-osmosis) water at the grocery store, but you also can use a carbon filter to prepare your water. **A** The downside to filtering your own water is that your pH and minerals can vary wildly based on your local water supply. See page 56 for more on brewing water, including more advanced topics such as pH and minerals.

The amount of sugar ... the mash that's dissolved in the wort you're boiling in your kettle is known as ... gravity (OG). It can be measured ... Brewers ... and homebrewers use a ... gravity scale, while American and ... brewers tend to use the Plato scale. The OG ... amount of alcohol content of your beer as well as the ... of your mash. A low-gravity session beer could be as low as 1.034 (SG)/8.5° Plato, while a strong imperial stout could be as high as 1.130/30° Plato.

The final gravity (FG) is the amount of sugar left after the yeast has had its ... The amount of sugar left ... on the ability of the yeast to turn the sugars ... all percentage of fermentable sugars in the wort. In general, the final ... is only about 15 to 20 percent of the original gravity. This can be helpful when deciding whether the beer has ... for bottling or not. If your beer's OG is 1.050, then the FG generally should be between 1.010 and 1.012.

Measuring gravity with a hydrometer is as easy as filling a container with liquid (at 60°F) and reading the ... Refractometer usage is equally easy to use. Just place a couple of drops on the surface and look through the eyepiece. Use a refractometer for original gravity only, because alcohol affects its reliability when it comes to reading final gravity.

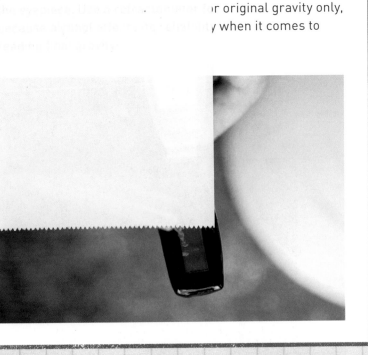

STEP 2 ➡ HEAT THE MASH WATER.

Using your stove or outdoor burner, heat your mash water to around 167°F (about 15°F higher than 152°F, which is a good temperature for your mash). If your grain is cooler than room temperature, you may want to start out 20°F higher. It's easier to add a small amount of cold water to the mash and bring the temperature down than it is to try to heat water (or the whole mash) if your temperature is too low.

STEP 3 ➡ MASH.

Mashing is simply mixing crushed grain and water together in order to activate enzymes that will convert the starch in the grain into sugars. There are simple mashes and complicated mashes. The vast majority of commercial breweries and homebrewers use a single-infusion mash, which holds the grain and water mixture between 148°F and 156°F for 30–60 minutes. See page 45 for more in-depth information on mashing.

Transfer the 167°F, to your mash tun (either a cooler or another pot). Mix the crushed grain and hot water together by holding the bucket or bag of grain in one arm and slowly adding it to the mash tun. **B** You can stir with the other hand or enlist the help of a friend to break up any lumps as you go. Once all the grain is added, stir the mixture well to ensure a consistent temperature throughout the mash. Put the lid on the cooler or pot and let the mixture sit for 45–60 minutes.

STEP 4 ➡ HEAT THE SPARGE WATER.

Sparging is the process of rinsing sugars off the grain after you have drained the initial mash water (the first runnings). Just as with mashing, there's a variety of sparging options. For at least your first few brews, I recommend fly sparging, which is what we'll do below. For more in-depth information on sparging, see page 54.

While you're waiting for the enzymes in the mash to do their work, heat your sparge water in your second kettle. Add the 5 gallons needed and heat the water to somewhere between 165°F and 170°F. This is a good temperature range because it will rinse a large percentage of sugars from the grain without the risk of extracting tannins that comes with using hotter water (more than 180°F).

STEP 5 ➡ RECIRCULATE (VORLAUF.)

After 45 to 60 minutes, your mash completes; additional time won't extract any additional sugars from the grain. If you can heat the mash, you can raise the mash temperature to 165–170°F. Known as a mashout, this allows the sugars to flow more freely from the grain bed. If you don't have the ability, don't worry; the effect is minor.

Attach a short piece of tubing to the spigot on your cooler or pot. Slowly open the spigot and recirculate, or *vorlauf* (the traditional German term), the wort back through the mash. Fill a large pitcher or bowl and gently pour the collected wort back on top of the mash. **C** This process helps you collect clearer wort and establish a good mash bed—which is key for a smooth runoff. Repeat the recirculation process for 5 minutes or until the wort is free from any grain particles. A little haze is okay.

STEP 6 ➡ COLLECT THE FIRST RUNNINGS.

Now you're ready to collect the wort. Start running the wort off into your kettle until the top of the grain bed is just dry. **D** Do this gently in order to minimize splashing.

If you're trying to shorten your brew day, start heating the kettle as soon as you have collected the first runnings. The wort will be close to a boil by the time you finish adding the sparge water.

STEP 7 ➡ SPARGE.

Now it's time to sparge, which will rinse the grain and let you collect second runnings. Gently pour 1 gallon or so of your sparge water on top of the mash until there are 2–3 inches of water above the grain. **E** Use a large spoon or similar item to diffuse the water. Stop pouring until the water drains and the grain bed shows, then repeat until you've collected 6 1/2 gallons of wort total (first and second runnings).

Note: If you have a refractometer, keep checking the gravity of the wort near the end. If it drops below 1.008, stop sparging to prevent the extraction of tannins from the grain husks.

B

C

D

E

STEP 8 ➡ BRING THE WORT TO A BOIL.

Once your first and second runnings are in your kettle, bring the wort to a boil. When the wort is near a boil, some protein and other assorted scum will float at the top. Feel free to skim that off.

STEP 9 ➡ BOIL.

Once the wort reaches a boil, let it boil for 15 minutes, then mix in the 60-minute addition of hops. **F** With 30 minutes left in the boil, add the other 7.5 alpha acid units of Columbus. Around the 15-minute mark is a good time to add your immersion chiller and any kettle finings (such as Irish moss or Whirlfloc) if you're using them.

STEP 10 ➡ CHILL.

At the end of the boil, turn off the heat source and start running cold water through the immersion chiller. **G** Throw in your flameout (knock-out) hop addition if your recipe has one. Depending on your water temperature, it probably will take between 20 and 40 minutes to cool your wort to pitching temperature. (For ales, usually around 65–70°F; for lagers, you may need to chill it to 65–70°F and then chill your beer overnight in a fridge or freezer to get it to proper pitching temperature.) Cover the wort with the lid of your pot or with a sturdy, clean trash bag as it's chilling.

STEP 11 ➡ TRANSFER AND CHECK GRAVITY.

When the wort has come down to pitching temperature, remove the wort chiller and transfer your wort to the sanitized fermentor using a sanitized racking cane or auto-siphon. **H** Pull about 8 ounces aside to check the gravity if you're using a hydrometer, or a few drops if you're using a refractometer.

STEP 12 ➡ AERATE AND PITCH YEAST.

Once the wort is in the fermentor, it's time to aerate the wort. Most brewers start by shaking the fermentor for about a minute; however, you may decide to infuse directly with oxygen. The latter is really necessary only for lagers or very strong ales. (See page 116 for more information on aeration.)

Once your wort is aerated, pitch your yeast as quickly as possible. **I** It's important to pitch the correct quantity of yeast for your beer. See page 110 for a chart of proper pitch rates for different types of recipes.

STEP 13 ➡ START FERMENTATION.

You can't do this step on your own; this is the yeast's job! But you should place your fermentor in an environment that's friendly for the yeast. For a typical ale, that means a dark, cool (65°F) room at the very least—if not an actual controlled fermentation chamber.

Most off-flavors from yeast are created as they reproduce during the first few days of fermentation. If you can control temperature for these few days, you greatly increase the chance of making great beer. See page 114 for more on yeast and fermentation temperature.

AFTER BREW DAY

STEP 1 ➡ 24 TO 48 HOURS

Check the blow-off tube or airlock for activity. If you're using a clear carboy, you'll see bubbles in the wort, too. As long as a few "blips" of air are coming through the airlock, there's nothing to worry about. As the 48-hour-and-beyond point approaches, though, you should see lots of air being forced from the airlock. If you don't, your seal is faulty or fermentation isn't taking off. If you're using a bucket, crack the lid and peak inside; if you see *krauesen* (foamy head) on top of the beer, then you're fine. In my experience, it's very rare for a beer not to start fermenting, so don't freak out.

STEP 2 ➡ ONE WEEK

The beer is just about done fermenting at this point, but it's still too yeasty to hurry it into bottles or a keg. Add any dry hops or specialty ingredients. Carefully crack the lid, toss in your hops or other ingredients, and quickly close the lid.

STEP 3 ➡ TWO TO THREE WEEKS

The beer completesly finished fermenting by this time and can be bottled or kegged. If you have a fermentation chamber, crash it down to 32°F and add a fining such as gelatin (page 123). After an additional 48 hours, you can transfer crystal-clear beer into your keg instead of cloudy beer.

PACKAGING

KEGGING

When I owned a homebrew shop, I always encouraged customers to make the jump to kegging. Over and over, I saw homebrewers who stopped brewing because of all the work (and mess) of bottling. Kegging—more than any other upgrade—will keep you brewing. It's hard not to dream about cleaning and sanitizing one big container instead of 50 small ones! Not to mention the instant gratification of trying a beer within an hour of putting it in the keg. Pouring your own beer from a tap in your own home is great!

Kegging Equipment

Kegs are nearly indestructible and will last a lifetime. They aren't as cheap as they once were, but they're still very affordable. Complete kegging systems can be purchased at your local homebrew shop or through many online suppliers (page 190). If you can source a carbon dioxide (CO_2) tank or kegs locally, you can save some money both on the items and on shipping.

Not covered on the following page is a spare fridge or chest freezer. It's possible to use a fridge for both food and beer, but it limits you to a single keg, and the real beauty of kegging is having multiple kegs on tap. (You also may have a partner who doesn't love kegs taking up all the room in the refrigerator.) Choosing whether to go with a fridge or freezer is a tough call. Freezers need a temperature regulator, which isn't cheap, and freezers also tend to have a shorter life span. However, you can fit more beer into a large chest freezer, and since you can set the temperature to anything you want with a regulator, you also can use it for lagering. Fridges last longer, but they keep the beer at whatever the temperature dial decides. A good place to look for either is on Craigslist or eBay, where you may find someone selling a fridge or freezer for cheap. That may help with your decision.

KEG J: The soda industry used to use 5-gallon kegs to dispense concentrated syrups. When they switched to the bag-in-a-box package, the market flooded with kegs. They could be had for as little as $5! Over the years, they've become more sought after and now sell for considerably more. You can purchase brand-new kegs, though their cost is typically double that of a used keg, and there's not a huge benefit in purchasing a new one.

The beauty of soda kegs (Cornelius or "corny" kegs) is that they'll last a lifetime and can be rebuilt with a new gasket set for just a few dollars. Also, most of these kegs hold exactly 5 gallons, which is perfect for homebrewers. Occasionally 3- and 10-gallon kegs pop up.

There are two varieties of soda kegs: pin lock and ball lock. These refer to the type of fittings that attach to the keg. Pin-lock kegs have connectors that snap over the in/out plugs on top of the keg, which have small pins sticking from them. Ball-lock kegs have smooth plugs and take a different fitting. Ball-lock kegs seem to be more prevalent in the marketplace, but both styles are perfectly fine for homebrewing. Since the disconnects are unique, it's easier if you commit to one or the other.

How many kegs do you need? Obviously that depends on how often you brew and how many beers you want on tap at a time. Since you don't want to have to empty your only keg before you can keg up your next batch, I recommend at least two kegs to start. You're going to want more eventually, so keep an eye out for deals.

FITTINGS AND TUBING K: You need a set of connectors that snap over the in and out plugs sticking from the top of the keg. One side is the CO_2-in connector, and the other is the beer-out connector.

The beer-out connector attaches to a dispensing tap with a length of tubing. The tap could be an inexpensive plastic squeeze tap or a gleaming copper bar tower. The CO_2 fitting connects to a regulator attached to your CO_2 tank via more tubing. Don't use siphon hose for kegging: You need thick-walled keg line, which is only slightly more expensive and available at any homebrew shop. I use a 2-foot length on the beer side with a plastic squeeze tap and a 4-foot length on the gas side. Luckily, you don't need a set of connectors for every beer on tap. You simply move them from one keg to the next.

CO_2 TANK L: Depending on how often you brew, you may be fine with a 5-pound CO_2 tank (which lasts for about 10 kegs). It's lightweight and easy to take to parties. A larger 20-pound CO_2 tank is more economical for the serious brewer. It's more of a pain to drag around, but it can last for more than a year between fills. (A perfect solution is to have both a large and small tank, but that's an added expense.) You can get the tanks filled or exchanged at your local welding supplier.

REGULATOR M: You need a regulator to step down the gas pressure to a usable level. A single-gauge regulator tells you how much pressure is going to the keg, which is your main concern. A double-gauge regulator has an additional gauge that tells you how much CO_2 is left in the tank. I've always found that the second gauge moves only right as the tank is running dry (usually in the middle of a party!). Weighing the tank is a better measurement of how much gas is left. A 5-pound tank weighs 5 pounds more when filled than when empty, and a 20-pound tank weighs 20 pounds more. Write the empty weight on the keg. Then all you have to do is place the tank on your scale to find out exactly how much CO_2 is left.

CHANGING THE GASKETS IN A USED KEG

Complete gasket sets usually are all you need to bring a used corny keg up to code. A set is inexpensive and consists of the large O-ring that fits around the lid, two small O-rings that fit over the plugs that stick from the top of the keg, and two tiny O-rings that fit over the dipstick tubes underneath the plugs. The hardest part is unscrewing the plugs. If you're lucky, you can get them off with a pair of pliers, but you may have to resort to a socket wrench to loosen them. A small amount of keg lube smeared on the new gaskets will help them seal and make it easier to remove the fittings later. Tighten the plugs well after replacing the dipstick-tube gaskets because it could be dangerous if they're loose. Occasionally, you'll run across a stubborn keg lid that just won't seal tight. The best option I've found is an extra-fat gasket sold by Williams Brewing (see References.) It's a few bucks more than a standard O-ring, but it works wonders. Once all the gaskets are replaced, your keg should seal properly, hold pressure, and work for years.

Kegging a Beer

STEP 1 ➡ CLEAN.

First, bleed any pressure from the keg. Lift the lid's relief valve. If your lid doesn't have a relief valve, carefully press down on the poppet of the gas side plug with a screwdriver. After venting, take off the lid and remove the large O-ring. Hose out any yeast or beer residue, then fill with hot water and about ¼ cup of PBW, OxiClean, or automatic dishwasher soap. Push down on both poppets with a screwdriver to allow the cleaning solution to fill the in and out dip tubes, loosely put the gasket on the lid, and submerge in the solution. After an hour-long soak, rinse the keg well with tap water. This job is best done outside with a hose. Make sure to spray water down the dip tubes by pressing down on the poppets again. If this is the first cleaning of a used keg, unscrew both plugs from the keg and replace the small gaskets under the dip tubes. Also replace the small O-rings on the plugs and the large lid O-ring.

STEP 2 ➡ SANITIZE.

Fill the keg with iodophor or the acid-based sanitizer of your choice. After the required contact time, close the keg lid and hook up the gas and liquid connectors. Push all the sanitizer from the keg using your CO_2. (You can flip the tap open so you don't have to squeeze it.) I push mine into a 5-gallon bucket and use it for sanitizing my siphon equipment. Why not just dump the sanitizer? This method purges all of the oxygen from the keg with CO_2 and sanitizes a liquid line in the process. When you siphon your beer into the keg, you don't have to worry about oxidizing your beer. This is especially important with IPAs and other beers with a nice hop aroma, because oxygen is the enemy of hop aroma.

STEP 3 ➡ FILL THE KEG.

Siphon the beer from your fermentor into the keg using a racking cane (or auto-siphon) and siphon tubing. Leave as much sediment behind as possible. If you didn't add a clarifier such as gelatin or Biofine® in the fermentor, add it to the keg at this time if you want.

STEP 4 ➡ SEAL AND PRESSURIZE THE KEG.

Close the lid and apply pressure with CO_2, which should seal the lid tightly. Set your CO_2 regulator to 30 pounds per square inch (psi) and place the keg on its side. Rock it back and forth, using your foot, for 30 seconds. Place the keg upright, unhook the CO_2 tank from the keg, and place the keg in your beer fridge or freezer.

After an hour or two, you can try your beer. Before tapping, however, you need to vent the excess pressure from the keg so it pours properly. Bleed off all the pressure by lifting the relief valve. Hook the CO_2 back to the keg with the pressure set to 0 on the regulator. Then turn your regulator to 10 psi, which is enough pressure to push the beer from the tap. Try pouring your beer!

If the beer comes out too fast, dial the pressure down until the beer pours properly. If the carbonation is a little too low, unhook the tap and pressurize the keg back up to 30 psi. This time, there's no need to shake it; just leave it hooked up for a few hours and check it again. It's easier to fix low carbonation than to fix over-carbonation, so take your time and work your way up to the desired carbonation. Always bleed off the excess pressure in the headspace of the keg before attempting to pour your beer. If it's shooting from the tap and foaming everywhere, then the pressure is too high! It should take at least five seconds to fill a standard pint glass.

Note: While this quick-shake method works well, brewers with a little more patience will set the CO_2 on their regulators to around 15 psi and leave them hooked up to the keg for five to seven days. This allows the beer to absorb CO_2 slowly, which means there's less chance of over-carbonating. The only downside (besides waiting a few days to try your beer) is the risk of losing all your CO_2 if there's a leak anywhere in the system.

Dispensing Beer from a Keg

Once your beer is carbonated, drinking it's as simple as setting your CO_2 regulator to a pressure that's low enough so that the beer flows from the tap at the speed you want. The pressure should keep it carbonated at the volume of CO_2 you want. The ideal pressure for your system will depend on the length and diameter of the tubing and the temperature of the beer. Somewhere in the 6–10 psi range works in most instances.

Note: If you want to go down a dispensing rabbit hole, the Brewer's Association Draught Beer Quality Manual (draughtquality.org) has more information than you'll ever need, including a chart to guide your psi settings precisely.

Since there's always the possibility of a small gas leak from a loose clamp or sticky poppet, keep the CO_2 turned off when you're not using it. If your beer becomes under-carbonated partway through the keg, just hook up the CO_2 tank again at 30 psi, fill the headspace of the keg, turn it off, and disconnect it. Leave the keg overnight to absorb the CO_2 charge, and the carbonation level should be back to where you want it the next day. This method would be completely impractical for a restaurant or bar, but at home it works just fine—and one day it will save you a tank of CO_2.

While dispensing can seem complicated —even overwhelming—at first, after a few times it will be so easy that you could do it in your sleep. Finding the right balance in your kegs is a bit of an art. Don't get too hung up on a single mistake. In most cases, the worst that can happen is a slightly over-carbonated beer. After a few days of bleeding off excess pressure, it will return to normal.

Filling Bottles from a Keg

Filling a bottle from your keg can be as easy as sticking the tap into the neck of a bottle and slowly filling it. You can get fancier by sticking your old plastic bottle filler into the squeeze tap so that you can fill from the bottom of the bottle (less splashing and less oxidation.) This technique is fine for taking beer to a party or a homebrew-club meeting, and it's the way I transport most of my beer. You don't even have to worry about sanitizing the equipment or the bottle. Since the beer is going to be consumed in a few hours, there's no worry about infections.

Beer filled in this cheap and easy method should last a week or so in the fridge, but for long-term storage or for competitions, you need to use a counter-pressure bottle filler. A counter-pressure filler allows you to purge the bottle with CO_2, pressurize the bottle to the same pressure as the keg, and then slowly fill the bottle. Since the bottle and the keg are at equilibrium, the beer foams very little and keeps most of its CO_2 in solution. When capping a bottle, have the beer foam up to the top of the beer neck. This is known to commercial brewers as "capping on foam" and ensures that the oxygen in the bottle is minimized. There's a variety of counter-pressure fillers on the market, and they all work decently. I prefer models that use a single valve to control the process, which leaves your other hand free to hold the bottle or grab a cap.

BOTTLING

Popping the cap on your first bottle of homebrew and hearing that familiar "hiss" are fun and exciting. Unfortunately, the fun and excitement quickly turn into a chore. Cleaning and sanitizing 50 bottles, a bottling bucket, and all your siphoning equipment take a lot of time. Then you have to store the beer for a couple of weeks before you can even taste it.

But don't write off bottle conditioning. You can get higher levels of CO_2 in bottled beer that are essential for certain beer styles, such as Belgian golden strong ale or hefeweizen. It's also best for really strong or sour beers because you'll want to enjoy these beers over a period of years and you don't want to tie up a keg for that long. There are also mysterious tastes that develop with bottle conditioning that you can't get with force carbonating, and the yeast can protect the beer from oxidation. (It can give "meaty" or "brothy" flavors if kept too long.)

Bottling Best Practices
STEP 1 ➡ DISCARD THE BAD BOTTLES.
Look into every bottle for crusty mold on the bottom or a ring around the bottleneck. If you see any, throw the bottle away—it's not worth the potential infection if you miss something. It takes a *lot* of time to clean a heavily soiled bottle.

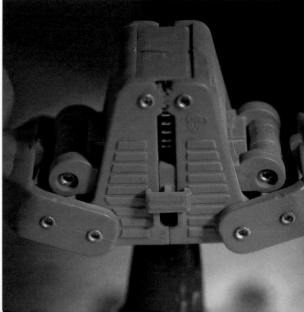

STEP 2 ➡ CLEAN AND SANITIZE THE GOOD BOTTLES.

Even if the bottles look clean, they should still get a quick soak in a cleaning solution of PBW or OxiClean. After rinsing, the bottles should be sanitized using iodophor or an acid-based sanitizer. Alternatively, the bottles can be loaded in an oven and baked for an hour at 350°F. Let the oven heat up and cool down slowly so that you don't shock the glass.

STEP 3 ➡ WATCH FOR DUST.

Cover bottles with foil to keep dust from falling into them. Do the same with the bottling bucket. Always be aware of the bacteria-laden dust all around us. A little foil over a cooling kettle or carboy goes a long way in preventing infections.

STEP 4 ➡ USE THE RIGHT AMOUNT OF PRIMING SUGAR AND MIX IT IN PROPERLY.

Under- and over-carbonated beers are common flaws. By following the priming sugar advice (at right), you can be sure of consistently getting the appropriate level of CO_2 in your beer—as long as it's mixed thoroughly with the beer. The most common rookie mistake is pouring the hot priming sugar mixture into the beer and starting to bottle. If you do that, you run the risk of ending up with flat bottles, some that are properly carbonated, and some that gush when you open them. If the priming sugar isn't mixed uniformly into the wort, the carbonation won't be uniform. Add the priming sugar mixture to the bottom of the bottling bucket or carboy, then siphon the beer on top of it. Give it a gentle stir with a sanitized spoon and proceed.

STEP 5 ➡ STICK WITH REGULAR CAPS.

While oxygen-absorbing caps are an option, keep in mind that they're activated by moisture. If you dip them in a sanitizing solution or boil them, their oxygen-scavenging capacity can be depleted in a matter of seconds. Also, new research shows that volatile hop aromas can be absorbed into the cap liners as well. Until the technology improves, save the money and buy the regular caps. The yeast in the bottle should be able to scavenge the small amount of oxygen in the headspace.

STEP 6 ➡ STORE THE BEER WARM, THEN STORE IT COLD.

Often a batch that won't carbonate has been kept too cool for the yeast to do its job. Belgian brewers have hot rooms for storing their carbonating beer. These rooms usually are in the 70–80°F range, and the beer sits there for about two weeks. This warm temperature encourages the yeast to consume the priming sugar and produce CO_2. Remember, the yeast cells are stressed out and tired at this point, so be nice to them! After two weeks of warm conditioning, chill a bottle and sample it. It should form a thick head and have the proper carbonation for the style. If not, shake the bottles a bit and let them sit for another week at the warm temperature. When carbonated, the beer should be stored cool or cold for the remainder of its life. Anywhere below 55°F is acceptable, but as close to 32°F as you can get will extend shelf life.

Bottle Conditioning: Sugar Ratios and Temperature

As a homebrewer, don't get too hung up on exact volumes of CO_2 in bottled beer—unless you really want to. At the end of the day, you just need to make sure that your beer forms a thick head when poured, but not to the point that the beer is half foam. That said, it's easiest to talk about CO_2 in volumes, so that's what we'll do here.

At my brewery, we focus on highly carbonated Belgian farmhouse-style beers and traditional English styles with low carbonation. Achieving the proper level of carbonation is very important because it can affect the mouthfeel and crispness of the beer. A saison may be bottled at 3.7 volumes and an imperial stout at only 2.3 volumes. Carbonation is an unappreciated "ingredient" that can make or break a beer. A flat saison or tripel is flabby and insipid, and an over-carbonated stout is harsh. The recipes in this book have the recommended amount of sugar for priming, but develop the habit of consulting an online calculator (do a web search for "priming sugar calculator") and tweak your carbonation for each batch you brew.

MALTS *and* MASHING

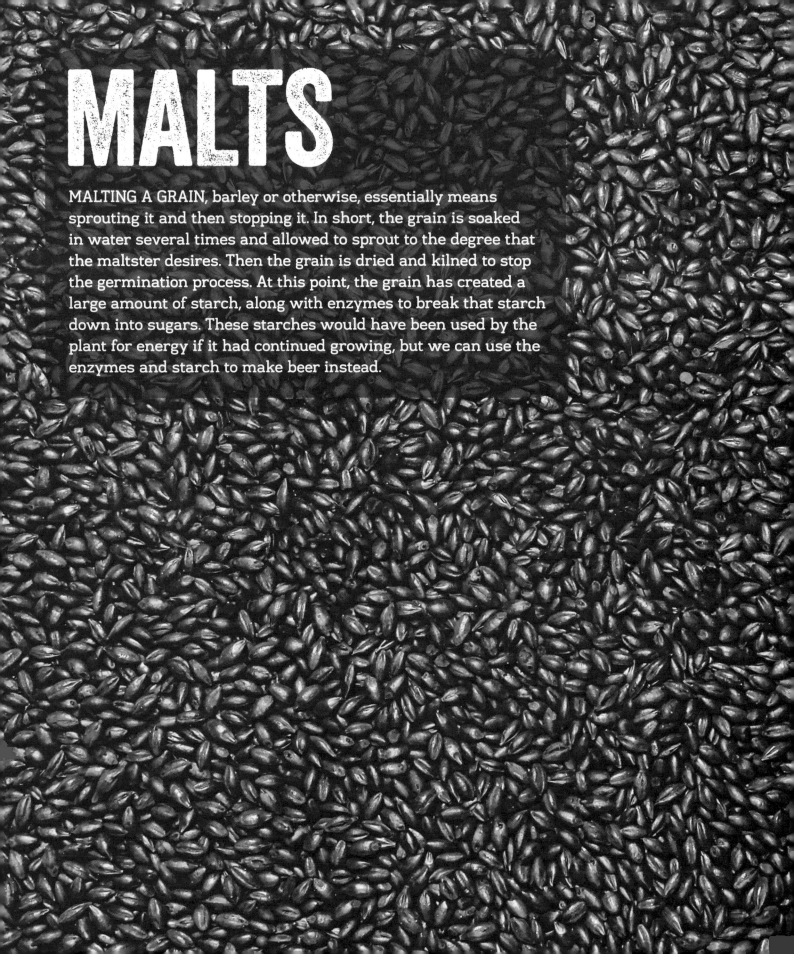

MALTS

MALTING A GRAIN, barley or otherwise, essentially means sprouting it and then stopping it. In short, the grain is soaked in water several times and allowed to sprout to the degree that the maltster desires. Then the grain is dried and kilned to stop the germination process. At this point, the grain has created a large amount of starch, along with enzymes to break that starch down into sugars. These starches would have been used by the plant for energy if it had continued growing, but we can use the enzymes and starch to make beer instead.

BASE MALTS

When brewers talk about malt, they're referring almost always to malted barley. Many other malted grains are available—oats, wheat, rye, spelt—but they're rarely used as the primary malt in a beer recipe.

Every brewing region has base malts tied to its brewing history, and each type of base malt has subtle variations in taste and appearance. The UK is well known for its pale-ale malt. Europe has pilsner, Vienna, and Munich malts. North America has its own traditional 2-row and 6-row malts. Over the last few decades, these boundaries have blurred: you can find German pale-ale malt, British pilsner malt, and U.S. Munich malt. It's a good idea, however, to go with the traditional regions for the type of malt you want. Nobody makes pilsner malt like the German maltsters, and nobody makes a pale-ale malt quite like a traditional British malt house.

These malts typically make up the majority of a recipe and provide the enzymes needed for the conversion of starch into sugar during the mash. All modern base malts have enough enzymes to convert themselves easily, and most have the power to help convert grains without enzymes (a.k.a. adjuncts) when they make up 15 percent or less of your grain bill.

Note: If you have a recipe that includes more than 20 percent starch adjuncts (rice, corn, pumpkin, etc.), it's a good idea to use North American 2-row or 6-row malt. These malts tend to have the most excess enzymes, which help convert the additional starches.

Base Malts at a Glance

Base Grain	Beer Style	Srm	Percentage Typically Used	Ability to Convert Other Grains
British Pale-Ale Malt	British Pale Ales, Porters, Barleywines, etc.	3–4	90–100%	low to medium
German pilsner	pilsner, Helles, Kolsch, etc.	1.5	70–100%	medium to high
German Vienna	Vienna lagers, Dunkels, or any style that needs a malt "bump"	4	10–100%	medium
German Munich	Munich Dunkels, Bocks, Dopplebocks	6–20	10–100%	low
American 2-Row	Any American style	1.5	90–100%	high
American 6-Row	"Macro"-style American lagers, any beer that uses a high percentage of adjunct	1.5	50–80%	very high
Malted Wheat	Wheat beers	1.5	10–70%	high
Malted Rye	Hoppy ales, Roggen beers	3.5	5–15%	medium

American

North American 2-row malt (1–2°L): If you want a crisp, clean malt taste without a lot of toasty, biscuit tastes, this is your go-to malt. The neutral taste also allows brewers to make strong beers (6–10% ABV) that are much more drinkable than an identical beer made with a British or European malt. There's not a lot of difference from maltster to maltster, but my favorite North American maltster is Great Western. Rahr also is an excellent choice.

Note: With 2-row, be careful when sparging your mash. If you oversparge or use water that's too hot or has a high pH on 2-row, you'll definitely extract a husky taste.

North American 6-row malt (1–2°L): This malt is similar to North American 2-row but with more protein and more husk, which can lead to hazy beer and an increased chance of picking up husky flavors. The higher levels of protein and husk make this malt useful for large breweries, which often use a hefty amount of rice or corn in their beers. The extra husks help keep the adjuncts in the mash from sticking, and the rice's or corn's lack of protein dilutes the protein. Just because large breweries use 6-row doesn't mean you shouldn't use it on a home-brew scale. In fact, 100 percent of 6-row lagers can taste great and score well in competitions. Don't be afraid of 6-row, but make sure to do a protein rest (page 48) and take the same precautions as 2-row when it comes to sparging.

British

British pale-ale malt (3–4°L): While many recipes simply call for "British pale-ale malt," many varieties are available, including Golden Promise, Halcyon, Maris Otter, Optic, and Pearl. Any variety will work fine in a British ale, but Maris Otter, with its strong, toasty malt flavor, is the most recognizable. Brewers differ in their opinions on Maris Otter. Some swear by it, while others don't think it makes much of a difference. I like the malt's flavor but find it can taste too toasty in some recipes. I'll often mix it with a less flavorful malt, such as North American 2-row. Golden Promise is another standout with which some brewers notice a marked difference. It's a Scottish malt with a distinctive sweet flavor, making it a good choice for Scottish ales (no surprise there).

If you start paying close attention and playing with a variety of British pale-ale malts, you'll soon come across floor-malted varieties. Floor malting indicates that the malt was turned manually while sprouting, as opposed to being turned by machines. It doesn't necessarily guarantee a better product, but it's more traditional and can serve as a sign of how much care a maltster is putting into the product.

German

German pilsner malt (1.5°L): The palest malt made in Europe, this malt is often used for 100 percent of a grain bill to make pilsners. It has a sweet flavor, with hints of graham crackers and honey. It's essential for re-creating European pale beers such as Kölsch, helles, and of course pilsner.

Vienna malt (4°L): Vienna is a slightly darker version of pilsner malt. When used at 100 percent of a grain bill, it will produce a deep, golden beer with a slightly more pronounced malt flavor. It's often used to beef up the malt flavor in amber lagers (and occasionally in ales) without adding too much color. Bell's® Two Hearted Ale® supposedly uses 10–15 percent Vienna, for example.

Munich malt (6–8°L): An even darker base malt than Vienna, Munich produces a light amber-colored beer with an intense malt aroma and taste when used at 100 percent. It's commonly used in the 10 percent range to add complexity to a large variety of beers, including IPAs and brown ales. Traditionally, it's an essential component in alt beers, dunkels, and bock beers, where it can make up all of the grain bill. Munich malt also comes in a dark version (15–20°L) that I prefer over the regular Munich malt. It has a flavor that just screams "German!" If you're substituting dark Munich into a recipe, use half as much as regular Munich malt and substitute pilsner malt for the rest. You also can make a beer with a heavy-handed use of dark Munich malt. Some of the best Munich Dunkel recipes use 50–60 percent dark Munich for the grain bill.

Beyond Barley

Malted wheat: Wheat gives a softer, rounder mouthfeel to beers and a pleasant grainy flavor. The protein in the wheat can give beer a haze, which is desirable in most wheat beers. It's an old homebrewer's belief that wheat enhances head retention, but this hasn't proven true in my experience. Beers with large amounts of wheat in them can have worse head retention than an all-barley beer. Wheat doesn't have a husk, so if you try to mash 100 percent wheat, you'll end up with a gummy mess. You can use rice hulls, which are just flavorless husks, in the mash to facilitate sparging or just keep the wheat at no more than 70 percent of your total grain bill. Raw wheat, which isn't malted, has a tangy, rustic flavor that's different from that of its malted brother. Other specialty wheats also are available. See the chart below.

Malted rye: Rye malt is hardly ever used in amounts greater than 15 percent except in a German Roggenbier. Some breweries tout the spicy flavor of rye, but malted rye does more to alter the mouthfeel of a beer than the flavor. It contributes an oily-feeling or "slick" quality that can be perceived as added body, which can be helpful in a session beer. Flaked rye doesn't seem to have this effect, and my experiments with using it up to 50 percent have shown that it has little flavor contribution compared with malted rye. Some maltsters are creating special types of malted rye along with specialty wheats. See the chart below.

Types of Wheat and Rye

VARIETY	WHEAT	RYE
1	**Malted Wheat**	**Malted Rye**
	Can be infusion mashed but benefits from a 15-minute protein rest.	Adds an oily, slick mouthfeel. Benefits from a beta-glucan rest (30 mins.).
2	**Raw Wheat**	**Raw Rye**
	Used in Belgian Wit and other styles. Needs to be crushed fine and preferably cereal mashed.	Only used in the form of rye flour that can be added to the mash. Benefits from a beta-glucan rest (30 mins.).
3	**Flaked Wheat**	**Flaked Rye**
	Can be infusion mashed, an easier form of raw wheat to work with.	Doesn't add the oily mouthfeel of malted rye. Little flavor contribution.
4	**Torrified Wheat**	**Fawcetts Caramel Rye**
	A "puffed" wheat usually used to lighten body in British ales.	Adds a tangy, caramel flavor to many styles.
5	**Weyermann® Dark Wheat**	**Weyermann Chocolate Rye**
	Used in Dunkel Weizens to add color and flavor.	Used in German Roggenbier. Typical chocolate malt flavor.
6	**Weyermann Chocolate Wheat**	
	A smooth dark wheat used in Dunkel Weizens but open for experimentation.	

CRYSTAL/CARAMEL MALTS

Crystal malts are base malts mashed to develop sugars, but because the grains are left whole, the sugars are trapped inside the grain. The sugar-filled grains then are kilned to caramelize the sugars inside. The flavors and colors produced from the kilning can range from barely notable to intense. It's hard to make generalizations about crystal malts because they vary wildly from maltster to maltster. North American and German crystal malts aren't as complex and don't give as much sweetness as others. British crystal malts are more complex and aggressive, whereas Belgian crystal malts fall somewhere between British and North American ones. The terms "caramel" and "crystal" are used for the same product, depending on where it was produced.

Light Crystal Malts
(10–30°L)

Examples: *British Light Crystal (sometimes called Caramalt or Carastan), North American Crystal (10–30°L), German Carahell (10°L), German Caravienne (25°L)*

American and German light crystal malts have a slight golden color but not a lot of caramel flavor or sweetness. I've made numerous pale ales with 30 percent light crystal malts, and the results have consistently underwhelmed. There's minimal sweetness or body contribution even when mashing at high temperatures. It may help to think of these malts as more like a base malt on par with Vienna when it comes to their character.

If you want a light crystal malt with more flavor, try substituting aromatic or melanoidin malt or try using British light crystal malts, such as Carastan or Caramalt, that contribute a delicate, toasty malt flavor without too much caramel character.

Medium Crystal Malts
(40–80°L)

Examples: *British Crystal 45–70°L (big difference between maltsters); North American Crystal 40–80°L (no real difference between maltsters); Belgian Caramunich; German Caramunich I, II, and III, (45, 60, and 75°L, respectively)*

Malts in this category tend to contribute a smooth caramel flavor and an amber color with usage in the 5–10 percent range. A handful of craft breweries, such as Rogue Ales, use large amounts of medium crystal malts in their beers, but most brewers feel that heavy-handed use of these malts muddies up the flavor and competes with the hops. It's a matter of personal preference, but if you find your beers are thick on the palate and your hop flavor seems muted, try easing up on the medium crystal malts. You can get your color with a small amount of darker malt instead.

Unlike with light crystal malts, selecting the right medium crystal for a recipe is crucial. British crystals tend to bring more raisin and burnt-sugar character and can vary quite a bit from maltster to maltster (and even batch to batch from the same maltster.) I like most British crystal malts in the 25–50°L range; they aren't too raisiny and add a lot of complexity without a ton of color. In contrast to the British malts, the German crystal malts are über-consistent but fairly restrained. I usually use only German caramel malts for German styles and the same for Belgian caramel malts and Belgian-style beers. North American medium crystal malts are similar to the German malts in that they're available in dozens of varieties that all have a crisp, clean flavor. Use them in pale ales, IPAs, and especially West Coast IPAs (where you don't want the crystal competing with the hops).

Dark Crystal malts
(120–160°L)

Examples: *Belgian Special B (150°L), Briess Extra Special Malt (135°L), Weyermann Caraaroma (130°L), Fawcett Dark Crystal (150°L)*

Dark crystal malts typically contribute strong raisin, toffee, and burnt-sugar aromas and tastes. They quickly can become overpowering if you use too much (no more than half a pound of any in a 5-gallon batch). Since most types of dark crystal have their own character, it's better to discuss the most popular ones individually.

Special B (150°L): Probably the most intense of the dark crystal malts, Special B has a strong raisin—almost winey—aroma and taste. It's made in Belgium and considered essential for many dark Trappist styles. In general, keep it below 5 percent.

Extra Special Malt (135°L): Made by Briess in America, Extra Special Malt is much milder than Special B and gives a complex flavor without any winey notes. Highland Brewing, a regional brewery on the East Coast, uses this malt as the only specialty malt in its tasty Gaelic Ale.

Caraaroma (130°L): Made by the Weyermann malt house in Bamberg, Germany, Caraaroma gives beers a beautiful ruby-red color and is similar in flavor to the Extra Special Malt above. As with any of these malts, half a pound in 5 gallons is plenty.

British Extra Dark (150°L): With lots of burnt-sugar and toffee character, this malt is great in red ales, extra special bitters (ESBs), or any beer that needs a deep mahogany color and lots of intense flavors.

CHARACTER MALTS

Character malts are like base malts on steroids. They pack a much more aggressive malt flavor. They can be used discreetly (5–10 percent of a grain bill) to add complexity, or they can go as high as half of the grain bill for a truly intense malt aroma and taste. Although this sounds great in theory, an overuse of these malts can produce cloyingly sweet beers, and the poor balance of flavors impedes the drinkability of the finished product.

Aromatic (25°L): This malt has an intense, almost exaggerated malt aroma and taste. At 5–10 percent, it's a great choice for bringing up the color and malt flavor of a beer without adding a caramel flavor. As the only character malt in an imperial IPA (at around 10 percent), it gives an attractive orange color and a complex malt flavor that isn't too sweet. Franco-Belges has a malt called "Special Aromatic," but it's really just a Vienna or Munich malt.

Melanoidin (23–30°L): Melanoidin is similar to aromatic but less intense. It was created to simulate the flavors from a decoction mash, which technically aren't melanoidins but Maillard reactions. It can be substituted for aromatic malt in most recipes.

Honey Malt (20°L): A unique malt from Gambrinus Malting in Canada, this tastes similar to honey, with an intense, sweet character. It's a no-brainer in a honey blonde recipe or in Belgian blonde or tripel. It can be quite intense for such a low Lovibond malt, so keep it under 5 percent to start with.

Biscuit/Victory®, Amber, and Special Roast (25°L): All of these malts are toasted malts that add a dry, nutty flavor to beer. Biscuit (Belgium), Amber (England), and Victory (U.S.) malts have a biscuity, peanut shell–like flavor. Special Roast is slightly darker than Biscuit and Amber at 55°L. It adds a tangy, almost sourdough aroma and taste. These malts can be used in the 5–10 percent range and are good malts to try in experimental pale ales.

Joseph Lovibond was a commercial brewer whose brewery existed until 1959. His lasting legacy, however, is his color scale. Invented in 1921, the scale is still the standard for commercial brewers today.

The palest malt you're likely to come across as a brewer will be about 1.5 degrees Lovibond (or 1.5°L). This would be an expected Lovibond rating for North American 2-row or 6-row malt and German pilsner malt. Malts continue up the scale to the blackest of the roasted malts at 550°L.

While the Lovibond rating conveys the color of malted barley and specialty malts, a scale known as the Standard Reference Method (SRM) denotes the color of the finished beer. The two scales are related: If you add the Lovibond ratings of the grain used and then divide by the number of gallons of beer, you get the SRM. For example, if you used 10 pounds of pale malt at 1.5°L plus ¾ pound of 60°L crystal in a 5-gallon batch, you would use this formula:

$$[(10 \times 1.5) + (0.75 \times 60)] \div 5 = 12 \text{ SRM}$$

Here are some color descriptions commonly used for beer and the corresponding SRMs:

Color	SRM
Light Straw	1 - 2.5
Pale Straw	2.3 - 3.5
Dark Straw	3.5 - 5.5
Light Amber	5.5 - 10
Pale Amber	10 - 18
Dark Amber or Copper	18 - 26
Very Dark Amber	26 - 40
Black	40 +

Europe uses a scale known as the European Brewing Convention (EBC). It's basically twice the Lovibond rating. (60 Lovibond is the same as 118 EBC.) When following someone else's recipe—especially someone from a foreign country—it's important to know which scale is being used. If a recipe from Germany calls for Crystal 65, then you need to know if that's Lovibond or EBC.

Roasted Malts

Roasted or dark malts usually fall between 300°L and 550°L—any darker than that and the grain would catch on fire! These malts contribute a dark color and tastes that range from chocolate to scorched toast. It may seem counterintuitive, but you can and often should use these malts to a higher percentage than crystal malts. For example, my house stout recipe uses 1½ pounds of black malt in a 5-gallon batch (page 63) to achieve a smooth dark-chocolate flavor. Let's start with the lightest and work up the list.

Brown malt (60°L): This is the original malt used to make traditional porter and stouts. A great secret ingredient in any dark ale, it gives a complex coffee and roast taste and varies greatly among maltsters. I really like Thomas Fawcett's version.

Coffee malt (165°L): This malt is similar to pale chocolate malt (below) but has an unmistakable coffee flavor. Brewers tend either to love it or hate it. Try it in a brown ale at 5 percent and see what you think.

Pale chocolate malt (215°L): This malt is a lighter version of chocolate malt, which is useful when you want to get some complex chocolaty flavors but don't want a

lot of color. It's a great choice for brown ales and ESBs. It's usually used at 3–5 percent.

Chocolate malt (about 350°L): This malt can add enough color to make a beer opaque without adding any burnt flavors. Unsurprisingly, it has a chocolaty aroma and taste that work well in any dark amber to black beer. Each maltster's chocolate malt has its own flavor, so definitely experiment. It's usually used in the 5–8 percent range.

Roasted barley (about 500°L): This is the black sheep of the specialty malts because it's the only one made from raw barley that hasn't been malted. This gives it a distinct flavor: slightly more ashy and burnt compared to black malt. It's traditionally used at around 10 percent in beers such as Irish-style stouts. Small amounts (2–3 percent) add color to Scottish ales as well.

Black malt (usually 400–550°L): Black malt is a fairly generic term that covers a huge range of malts with wildly different flavors. This category used to be called "black patent malt," and that name is still kicking around—especially for British malts. Black malts can range from smooth, bittersweet chocolate flavors to charcoal. The only way to get to know these different malts is to brew with them. Develop a simple house stout recipe with around 10 percent black malt and repeatedly substitute different black malts every time you brew. With a good base recipe, it always will be tasty, and it always will be different. Here are a few black malts that stand out:

➡ **British black malt:** This one has a wonderful smooth, bittersweet chocolate flavor but doesn't bring any harsh character to the beer. It's my go-to malt for American black ale and roasty stouts, and I usually use Crisp or Thomas Fawcett brand.

➡ **Dehusked Carafa malt:** Removing the husk minimizes any chance of harsh bitterness. This malt has a more refined flavor than British black malt, but it sacrifices complexity in order to get it. It's a great malt to use when making dark German or Czech lagers, Baltic porters, and black IPAs. It also can be used in small amounts to get a hint of color in many amber beers. It comes in three degrees of roast: Carafa I, II, and III (330, 410, and 525°L, respectively). The Carafa name is a trademark of Weyermann, but other manufacturers use the terms "debittered" or "dehusked" to designate a similar product.

➡ **Midnight Wheat®:** A black wheat malt made by Briess that's similar to the Carafa malts, Midnight Wheat has no husk because it's wheat. It's unusual in that you can use it to make a pitch-black beer with virtually no roast flavor. That feature makes it a great choice for black IPAs or Belgian dark ales, where you want a strong ratio of color contribution to flavor.

➡ **North American black malts:** These usually are made with 6-row malt and have a very intense and distinct flavor. I would describe them as having dark molasses and licorice taste at normal usage (5–10 percent). At higher rates, the flavors become harsh and almost phenolic (plastic-like aroma).

OTHER MALTS
Smoked Malts

Back in the day, all grain was dried after sprouting by using wood smoke (or perhaps peat in Scotland), so all beer had a bit of smoke flavor. Nowadays, smoked beer is making a comeback among craft brewers, especially with smoked porters. Some brewers use pre-smoked malt, which is available from several maltsters, and some prefer to smoke their own. Smoking a large quantity of malt is a huge endeavor, but for homebrewers who need only a few pounds, it's more manageable.

Peat-smoked malt: This malt from Scotland, which is used to make scotch, has a phenolic aroma that most drinkers find objectionable. Some brewers continue to use it. I would never use more than 1–2 percent in a Scottish ale, and I strongly recommend starting with another smoked malt for better aroma and taste.

Beechwood smoked malt: This German malt is used to make the famous rauchbiers in Bamberg. These malts often are used for up to 90 percent of the recipe, but I would start at 40–50 percent and work up from there. Drinking 5 gallons of an intensely smoky beer can become a chore!

Other smoked malts: Briess makes an intense cherrywood-smoked malt that gives an intense, smoky flavor to a beer when used as little as 5–10 percent. Weyermann Malt also makes a special wheat malt smoked using oak instead of the beechwood they use to smoke their barley. Like the German smoked barley, it can be used as up to 90 percent of the mash, but keep it at 70 percent or lower.

Acid Malt

Its own naturally occurring lactic acid sours this malt from Germany. It traditionally was used to lower the pH of the mash by brewers who followed Reinheitsgebot, which forbids using anything other than malt, hops, yeast, and water when brewing beer. Brewers not restricted by outdated Bavarian brewing laws can add lactic or phosphoric acid to lower the pH of their mash if necessary. (See page 56 for more on pH.)

Although it's tempting to throw several pounds of acid malt right into your mash for an easy sour beer, you need to remember that the enzymes in malt need a particular pH in order to convert the starch in the malt. If you add enough acid malt to make the mash sour, then you're also making the mash too acidic for the enzymes to do what they need to do. If you really want to go this route, adding lactic acid to taste in the finished beer is a better solution.

TOP MASHING AND COLD STEEPING

Brewers occasionally want to add color to a beer but keep the flavor impact more subdued than a normal mash. (Typically they're trying to avoid the deep roasty flavors that usually come with the darker malts.) Here are three ways to add color and subdued flavor with dark malts.

➡ **Top mashing** is a simple technique: You wait until your main mash completes, sprinkle the dark malts on top of the mash, then begin sparging. This minimizes the contact time of the dark malts and results in a smoother, mellower flavor while still extracting color. This was a popular procedure among homebrewers before the new dehusked black malts were available, and it remains a useful technique.

➡ **Cold steeping** is a method taken from the coffee industry. You mix the crushed dark malt with enough cold water to cover it and let it sit overnight. Then strain the mixture through a coffee filter into a jar. It's usually added to the kettle at the end of the boil. The advantage of this method is that you can add just enough of the tea to get the color you want with minimum flavor contribution.

➡ **Sinimar®** is a commercial malt extract made entirely of black malt. You can add it at any time, from the kettle to the kegged beer. It's a great way to make color adjustments late in the game or even to change a beer into a different style. Use it to turn your hefeweizen into a dunkelweizen in a matter of minutes.

TOASTING SPECIALTY MALTS

Modern specialty malt manufacturers can get a vast array of colors and flavors that are consistent from batch to batch. Toasting malt at home in your oven can yield good results, but it can be pretty inconsistent from batch to batch. You might wonder why you'd want to make your own. My answer is that it's still fun—and you learn about the toasting process. Even if the toasting schedule isn't exactly repeatable, there's still something to be said for starting with a sack of pale malt and being able to make a variety of specialty malts from it.

YOU NEED

oven

baking sheet

aluminum foil

2–3 pounds North American or
    British 2-row malt

paper bag

Making Toasted Malt

The simplest of specialty malts to make at home is a toasted malt. A lightly toasted pale malt would be in the 20°L range and would be similar to Biscuit malt. If you keep heating the grain, you get into the Amber malt or Special Roast–type malts. These give more of a red hue and a distinct bready flavor to a beer. You can even wait until the malt starts to turn a light-brown color and create a brown malt with an almost coffee-like character.

1. Preheat your oven to 350°F.

2. Cover your baking sheet with foil and place malt on top. (The foil makes removing the grain from the pan much easier.)

3. Place the sheet in the oven. Toast for 20 minutes, stirring occasionally, for malt in the 20°L range. Toast for 30–40 minutes for malt in the 40–60°L range. For an even darker malt, along the line of an 80°L brown malt, turn the oven up to 400°F and leave the malt in for

almost an hour. Stir it occasionally, but stir it very frequently near the end.

4. Remove the sheet from the oven and let the malt cool.

5. After the malt is cool, fold the foil into a makeshift funnel and pour the malt into a paper bag. Most people who regularly home-toast recommend letting the malt sit for a few weeks before using it. This curing process allows the grain to release some volatile flavors.

6. After it rests, spotlight your malt in a beer so you can taste the results! A simple amber ale with a clean base-malt flavor and not too much going on in the hop department works. See page 40 for my recipe.

Making Crystal/Caramel Malts

Making crystal malt is similar to making toasted malt except that you need to develop the sugars in the grain before toasting it. This is the same technique as mashing, but the grain isn't crushed first, and you aren't going to try to rinse the sugars off the grain afterward. Instead, you want the starch inside the grain to convert into sugar. Then you caramelize those sugars by heating them up.

In addition to the supplies needed for toasted malts, at left, you'll also need a large pot to hold 2 pounds of uncrushed pale malt, filtered water, and a thermometer.

1. In a pot, soak 2 pounds of uncrushed pale malt in room-temperature filtered water (enough to barely cover the grain).

2. Heat the mixture to 155–160°F and hold it for another hour.

3. Preheat the oven to 250°F. Drain the grain and spread it on a baking sheet as with the toasted malt. The malt needs to be dried before toasting, so put it in the preheated oven for about 2 hours.

4. When the malt is dry, you could stop and have a light crystal malt around 10°L. Turning up the oven to 350°F and waiting 15–20 minutes will give you a medium crystal in the 40–60°L range, and waiting for 45 minutes will give you a full-flavored dark crystal malt in the 100–120°L range. As with the toasted malts, crystal malts should age in a paper bag to mellow for at least two weeks.

Making Roasted Malts

Making roasted malts is very tricky without the right equipment. In a malt house, a roaster activates a water spray to chill the grain quickly right before it catches fire. Oh yeah—there's a lot of smoke involved. You really don't want to do this in your kitchen! But if you're in a postapocalyptic scenario and you need to make stout from a sack of pale malt, here's what you should do.

1. Preheat your oven to 450°F.

2. Start out like you're making toasted malt, stirring the malt often for about an hour.

3. When the malt starts turning chocolate brown, keep a close eye on it. When you smell the slightest hint of smoke, take the pan from the oven and put it outside. Wrap the grain up in foil and let it cool.

4. Store it in a paper bag for about two weeks to mellow.

5. You'll have to judge by taste how much you want to use in a recipe: 5 percent should give you a smooth porter or brown ale, while 10 percent should give you an almost black stout. Even if the beer comes out great, you'll appreciate being able to buy roasted malt at your local homebrew shop from now on!

DIY AMBER

Since this beer is designed as a showcase for your home-toasted malt (page 38),
the color and flavor depend entirely on how dark you go with your toasting.
It may be deep gold with a delicate biscuit flavor or dark brown with roasted flavors.
The hop bitterness stays low so it doesn't interfere with the malt.

YOU NEED

basic brewing equipment (page 3)

8 ½ gallons filtered brewing water (page 12)

9 pounds pale malt (81.8%)

2 pounds home-toasted malt (page 38) (18.2%)

7 alpha acid units Willamette hops at 60 minutes (25 IBU)

1 Whirlfloc tablet

2 vials or packages California Ale WLP001/ American Ale WY1056 yeast (or a 2-liter starter made from 1 pack; page 110)

3¾ ounces dextrose/corn sugar (optional, use only for bottling)

TARGETS
Yield: 5 gallons
OG: 1.053–1.055
FG: 1.012
IBU: 25

1. Mix the malt with 3½ gallons of water at 170°F or the appropriate temperature to mash at 155°F. Mash for 60 minutes.

2. Recirculate the wort until it's fairly clear. Run off the wort into the kettle.

3. Sparge with 5 more gallons of water at 165°F. Run off the wort into the kettle.

4. Bring the wort to a boil. Boil it for 15 minutes, then add the hops and continue to boil for 60 minutes. Add the Whirlfloc tablet at 30 minutes. Put your wort chiller into the wort at least 15 minutes before the end of the boil.

5. When the boil finishes, cover the pot with a lid or a new trash bag and chill to 65°F. Siphon the wort into your sanitized fermentor and pitch two packs of liquid yeast or a 2-liter starter.

6. Ferment at 65°F for one week. During the second week, let the temperature rise to 68–70°F.

7. Keg or bottle the beer. If you're bottling, use 3¾ ounces of dextrose/corn sugar for this beer.

SMaSH IPA

The only way to know exactly how a specific hop or malt tastes is to brew a beer using the hop or malt in isolation. A Single Malt and Single Hop (SMaSH) beer kills two birds with one stone—it gives the spotlight to one variety of malt and one type of hop. The overall character of the beer can vary greatly depending on your choices. A beer made with Munich malt and Centennial hops will be a different animal than one brewed with floor-malted Maris Otter and Fuggles. Make sure to take good tasting notes.

YOU NEED
basic brewing equipment (page 3)

9 gallons filtered brewing water (page 12)

13 pounds base malt (your choice, 100%)

15 alpha acid units hop of your choice at 60 minutes (54 IBU)

1 Whirlfloc tablet

1 ounce same hop at 15 minutes (10–20 IBU)

3 ounces same hop at end of boil (0 IBU)

2 vials or packages California Ale WLP001/ American Ale WY1056 yeast (or a 2-liter starter made from 1 pack, page 110

3 ounces same hop after primary fermentation (dry hop)

4.4 ounces dextrose or corn sugar (optional, use only for bottling)

TARGETS
Yield: 5 gallons
OG: 1.060–1.064
FG: 1.010
IBU: Around 70

1. Mix the malt with 4 gallons of water at 165°F or the appropriate temperature to mash at 150°F. Mash for 60 minutes.

2. Recirculate the wort until it's fairly clear. Run off the wort into the kettle.

3. Sparge with 5 more gallons of water at 165°F. Run off the wort into the kettle.

4. Bring the wort to a boil. Boil it for 15 minutes, then add the hops and continue to boil for 60 minutes. Add the Whirlfloc tablet at 30 minutes. Put your wort chiller into the wort at least 15 minutes before the end of the boil. Put in the other hop additions.

5. When the boil finishes, cover the pot with a lid or a new trash bag and chill to 65°F. Siphon the wort into your sanitized fermentor and pitch two packs of liquid yeast or a 2-liter starter.

6. Ferment at 65°F for one week, then let it warm to 68–70°F for the second week. Add the dry hops after primary fermentation slows down, usually after 5 to 7 days. Keg or bottle the beer. If bottling, use 4.4 ounces of dextrose or corn sugar for this beer.

MASHING, SPARGING, AND WATER ADJUSTMENT

MASHING

Mashing is simply mixing grain and hot water together in order to let the enzymes in the malt convert the starches into sugars. The temperature of the mash determines to some extent the fermentability of the wort and, therefore, the body of the finished beer. The length of a mash can be as short as 20 minutes or as long as several hours, although 60 minutes is commonly accepted as an adequate amount of time. There's little difference between mashing on a commercial scale and homebrewing except that professional brewers have a grist hydrator, a tool that sprays the water onto the grain as it enters the mash tun. Some commercial breweries have mash rakes that rotate in the mash tun to mix the grain and water together, but others use a paddle and manpower. The mash tuns in commercial breweries have slotted bottoms—commonly known as false bottoms—that allow the wort to separate from the grain. Home-brewers have invented a myriad of straining mechanisms (page 4).

The Temperature Zones

For some of the various mashing options that follow, it helps to understand a bit about the effect that certain temperature ranges have on malt. Many brewers don't know that as soon as malt meets water, things are going on at an enzymatic level—no matter what the temperature! Even at room temperature, starch begins hydrating and slowly converting, and the pH of the mash slowly drops. Old German brewmeisters called this an **acid rest**, but for modern brewers it's little more than a historical tidbit. Modern mashing doesn't require an acid rest, and even when a commercial brewer wants to adjust the pH, he or she typically uses acid malt or food-grade acid instead of an acid rest.

Raising the mash temperature to between 115° and 117°F is a rarely used rest known as a **beta-glucan rest**. Brewers find it useful when their recipe contains large amounts of gluey malts or adjuncts such as rye or oats. This rest will help break down the sticky beta-glucans (long-chain, unfermentable sugars) and hopefully make the mash and sparge more manageable.

Raising the mash temperature to between 122° and 135°F is known as a **protein rest**. This temperature zone energizes the proteolytic enzymes that will break up the proteins that can cause haze in a finished beer. These enzymes also can impact head retention and mouthfeel adversely, so a limited amount of time should be spent at this range. On the rare occasions that I do a protein rest, I aim for the higher end of the range (around 132°F) and hold it for just 10 to 15 minutes.

Once the mash is between 140° and 148°F, you're at the **low end of the saccharification rest**. In this range, diastatic enzymes known as beta amylases kick-start and begin to make short-chain sugars from the starch, which are highly fermentable. The longer the mash stays in this temperature range, the crisper and drier your finished beer will taste. Raise the mash temperature above 150°F for at least 10 to 15 minutes after this rest to ensure that all the starch in the barley is gelatinized and available to the enzymes.

Note: To gelatinize a grain means that you are getting it to absorb as much water as it possibly can. Because enzymes can't work on starches that haven't been gelatinized, it's important that any starch you use is fully gelatinized. Since grains don't absorb much cold water, hot water is required for complete gelatinization. How hot the water needs to be depends on the particular starch and how finely it's ground. Crushed barley is gelatinized at normal mash temperatures, but other grains should be boiled for up to an hour (page 161).

The next step up is from 149° to 160°F. This is the **high end of the saccharification rest** and the range most brewers mean when they refer to a saccharification rest. This rest turbo-charges enzymes known as alpha amylases. These enzymes create longer-chain sugars and dextrins from the starch, which are less fermentable than the short-chain sugars and contribute to the mouthfeel of the beer. This alpha-amylase zone is typically the only temperature range used by modern brewers. It's the optimal temperature in which many enzymes work their magic and completely convert nearly all of the starch in the mash to sugar in as few as 15 minutes. However, 45- to 60-minute mashes are more common.

The final temperature range, when the entire mash temperature is raised to 170°F, is often referred to as **mashout**. This temperature facilitates the runoff of the wort and deactivates the majority of the enzymes. It helps lock down the fermentability of the wort.

Commercial brewers rarely use it, mostly because of equipment limitations, but homebrewers commonly use it—especially when mashing grains that can cause a stuck mash, such as rye or wheat.

Extraction Rates and Evaporation Rates

Points per pound per gallon (PPG) is the amount of sugar (in gravity units) that 1 pound of grain or sugar will add to 1 gallon of water. For example, 1 pound of cane sugar in 1 gallon of water gives you a gravity of 1.046 or 46 PPG. Dividing by 5 will give you the gravity for 5 gallons: 46 ÷ 5 = 1.009.

Different grains yield different PPG. Base malts such as pilsner, pale ale, or 2-row have the potential to give you 37 PPG, but that's in a laboratory setting. Most homebrewers get 65–75 percent of the potential extract, which is called your "efficiency." To find out the efficiency of your particular brew system, mash 10 pounds of base malt and collect 6 gallons of wort. The theoretical maximum is 10 (pounds) × 37 (PPG) ÷ 6 (gallons) = 61 or 1.061. If the actual gravity of your wort is 1.046, then your brewhouse efficiency is 46 ÷ 61, or 75 percent.

Once you know your efficiency, you can determine the gravity of your wort fairly accurately based on the grain used. It gets a little tricky when you start throwing in different specialty malts because they have varying PPG depending of the level of roasting. Here are some typical PPG for various malts with a standard efficiency of 75 percent:

Base malts (2-row, pilsner, pale ale)	28 PPG
Dark base malts (Munich, Vienna)	26 PPG
Crystal malts	25 PPG
Chocolate malts (300–400°L)	21 PPG
Black malts (400°L+)	19 PPG

Raw grains can vary wildly in PPG depending on gelatinization temperatures and on how finely they're milled. See "Raw Wheat Experiment," page 160.

With this information you can calculate the gravity of a 6-gallon batch of porter with the following recipe:

9 pounds pale malt
1 pound Munich malt
1 pound crystal malt
½ pound chocolate malt
½ pound black malt

The PPG of each malt above is easy to figure out.

Pale malt: 9 x 28 (PPG) ÷ 6 = 42
Munich malt: 1 x 26 ÷ 6 = 4.3
Crystal malt: 1 x 25 ÷ 6 = 4.2
Chocolate malt: 0.5 x 21 ÷ 6 = 1¾
Black malt: 0.5 x 19 ÷ 6 = 1.6

Adding these together gives the PPG for the porter: 54 or 1.054.

Of course, this is your pre-boil gravity. Your gravity will increase as you evaporate the water and concentrate the wort. Usually you'll concentrate the wort from 6 gallons to 5 gallons, which will give you a starting gravity of 1.065 in this scenario. (Divide the above formulas by 5 instead of 6 to get the post-boil numbers.)

Evaporation rates can be difficult to predict, however, because they depend on the intensity of the boil, the surface area of the kettle, and even humidity. Commercial brewers talk about evaporation as a percentage. When they say they get 10 percent evaporation, it means that they lose 10 gallons for every 100 gallons during a typical boil. However, homebrewers tend to think about evaporation in terms of gallons. If you typically lose 1 gallon for every 60 minutes of the boil (or 1½ gallons during a 90-minute boil), that means you'll need 6½ gallons of wort in the kettle for a 60-minute boil or 7 gallons for a 90-minute boil to end up with 5½ gallons at the end. After a few brews, you'll know how much typically evaporates and collect enough wort so you end up with the desired volume at the end of the boil.

CRUSHING IT

Theoretically the perfect crush would be almost completely intact grain husks, with the inside of each husk crushed into flour. This would allow easy and complete gelatinization and saccharification of the starch flour while ensuring a good filter bed from all the large husk material. In practice, this is almost impossible to achieve. If you adjust your grain mill with too tight a gap, then the husks will tear to bits—possibly resulting in a husky, astringent flavor and a stuck mash. If you adjust your mill with too much of a gap, then you will have uncrushed kernels, and your efficiency will drop like a rock. The happy medium that brewers want is to have no uncrushed grains and a good amount of large pieces of husk to form a good filter bed. If you notice overly low efficiency from your mash (less than 65%), then try a tighter crush on your malt. Sometimes running the malt through the mill a second time also can help. Of course, if you start having stuck mashes or a harsh, tannic flavor in your beer, then you may want to crush a little more gently and just add an extra pound or two of grain.

Note: The grain on the left has been crushed too finely. Notice how the husks aren't as intact as those in the piles to the right. The crush on the far right has gone too far in the other direction. You can see whole pieces of barley that slipped through the mill uncrushed. The ideal crush is the one in the middle.

MASHING OPTIONS

How does a brewer decide what kind of mash regimen to use? For one, he or she may be limited by equipment. A plastic cooler doesn't allow for multiple steps without adding large amounts of boiling water. The grain bill might make a protein rest undesirable (all British malt), optional (German malts), or necessary (lots of rye/oats). In this section, we'll go over the various mashing methods and when to use them.

Single-Infusion Mash

Modern commercial brewers rarely concern themselves with anything other than an infusion mash at a single temperature somewhere between 146°F and 160°F. As I mentioned in the "The Temperature Zones" (page 45), the lower the temperature, the more fermentable the wort will be; the higher the temperature, the less fermentable it will be. Most breweries, however, settle on a happy medium for most brews and mash at 150–153°F for 45 to 60 minutes. This suits modern brewing systems, which rarely allow the heating of the mash (and it avoids the financial increases in energy and time required to perform multiple-step mashes).

Here's an example of the type of mash that the majority of commercial breweries use: a simple infusion mash that uses strike water around 15°F hotter than the desired mash temperature (20°F if the grain is cold). The grain absorbs some of the heat, which drops the mash temperature into the desired range. This works extremely well with British malts and also is appropriate for most North American and European malts.

Infusion Mash with a Protein Rest

If a brewery is doing anything other than a basic infusion mash, chances are they're doing a two-step mash so they can utilize a protein rest. Basically, this means mashing in at a lower temperature (122–135°F) and resting there for 10 to 20 minutes before ramping up to the conversion temperature range. Because most modern malt has had its protein degraded during the malting process, the overwhelming majority of commercial craft breweries would rather save the money and time and skip this step, but it still has its place for some breweries and in some

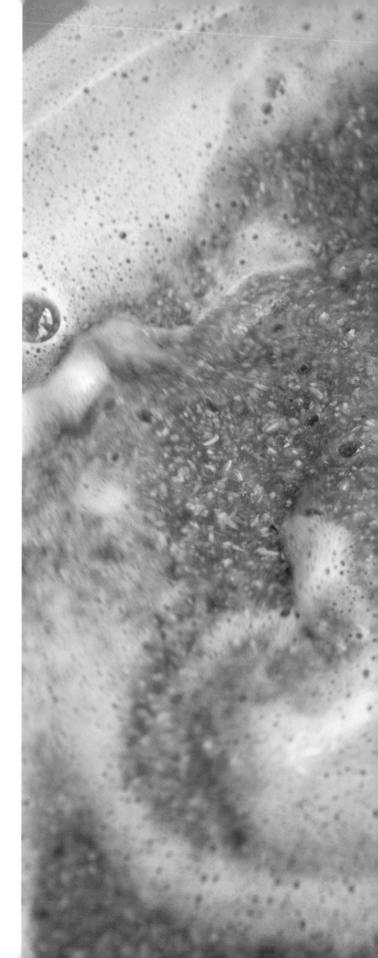

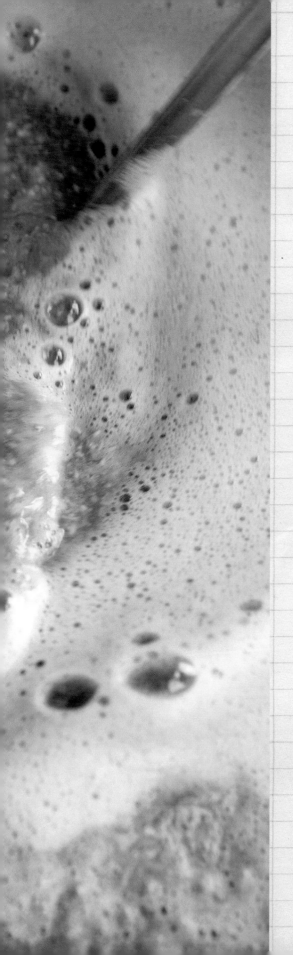

MASH TEMPERATURE EXPERIMENT

Common brewing knowledge tells us that mashing at a higher temperature will give us more unfermentable sugars and, therefore, a fuller body, while a lower-temperature mash should yield more fermentable sugars, consequently giving us a drier, crisper beer.

For this experiment, we tested that hypothesis with two identical mashes. Batch 1 was mashed at 146°F and Batch 2 was mashed at 164°F. The recipe was:

4 pounds British pale-ale malt
4 pounds German pilsner malt
1 pound German Caramunich II
8 alpha acid units Fuggle at start of boil

The original gravity of both beers was exactly 1.044, so there was no difference in extraction. It was pitched with California Ale WLP001 yeast and fermented at 68°F for 10 days.

Batch #1 (146°F mash) final gravity: 1.006
Batch #2 (164°F mash) final gravity: 1.016

A difference of .010 in final gravity is pretty substantial. The higher gravity should result in more body and sweetness, but surprisingly the opposite seemed to be true! I gave a side-by-side blind tasting to 10 professional brewers and beer judges, and nine of them chose the low-temperature mash as having more body (although just slightly). This flies in the face of conventional wisdom and implies that final gravity has little to no impact on perceived body.

I assume that the extra alcohol created in the low-temperature mash gives the impression of sweetness and body. What I take away from this experiment is that the only reason to mash at high temperatures is to limit the amount of alcohol in a session beer. Otherwise, keep those mash temperatures in the 148–152°F range.

recipes. Many if not most old-school European breweries continue to use a protein rest—and who can argue with tradition? I perform a 15-minute rest at 132°F whenever I'm using North American 6-row malt or European pilsner malt and am concerned about beer clarity. It's worth noting, though, that this slight increase in clarity has to be weighed against a potential decrease in body and head retention, although this is more likely with a rest of 30 minutes or longer. It depends on the malt: You want to break the excessive protein down to an appropriate level but not to eliminate it completely. Some malts might take 5 minutes; others may take 20 minutes or more.

Decoction Mash

Before the invention of thermometers, brewers figured out that if they mashed in at room temperature and then kept taking a portion of the mash, boiling it, and adding it back to the main mash (known as a "decoction"), they would get better results. Different brewers performed single, double, or even triple decoctions. (Talk about a long brew day!) Looking at it with a modern eye, we see that they likely were hitting the classic temperature ranges: 122°F, 145°F, and 162°F. But another benefit was the flavors created by the Maillard reaction, which happens when proteins encounter high heat.

Some brewers claim that the flavor contribution is negligible, while others swear they can taste the difference between a decoction-mashed beer and a standard mash. Regardless of whether there's any notable difference with decoction mashing, I still enjoy employing it on ultra-traditional styles such as Munich Dunkel or traditional bocks and dopplebocks. I enjoy the connection with the brewers of the past and traditional ways of brewing.

Key Temperatures for Decoction Mashing

Room temperature: This is mostly for historical purposes. It can take several hours to lower the pH of the mash to desired levels.

117–120°F: This range is commonly used for German hefeweizens, but only if you want to increase the spicy clove-like character. It also can be used for a grain that has a lot of beta-glucan in it (such as rye.) Rest it for about 30 minutes.

132–135°F: Mashing in this range may help with clarity by getting rid of chill haze, but it also also may hurt head retention and body. The ideal amount of time to keep it here is about 10–15 minutes. Never hold it for longer than 20 minutes, if possible.

146–162°F: This is the range required for the conversion of starches. With most modern malts, aim for the low end for a more fermentable wort and for the high end for less fermentable wort. Rest for 60 minutes for complete starch conversion.

Kesselmaishe (My Favorite Decoction Mash)

When I want to make a beer with a decoction mash, I turn to an old style known as Kesselmaishe (German for "kettle mash"). The beauty of Kesselmaishe is that it allows you to go through a standard mashing schedule, drain off the enzyme-rich wort, boil the entire mash, and then add the wort back for a final conversion and mashout. In other decoction techniques, you typically must stay at certain rests for longer than you would like as you go through the process of boiling the decoction. This results in a protein rest of 30 to 45 minutes, which isn't desirable. Kesselmaishe not only protects the wort that's full of enzymes, but it also boils the full mash—which results in maximum Maillard reactions and, therefore, maximum flavor contributions from the added work you're doing! If you want to perform a decoction mash for fun, nostalgia, or flavor, this is the technique I recommend.

1. Conduct the mash as you normally would, including a protein rest if desired. After you have reached the saccharification rest (146–155°F), wait 30 minutes, then drain all the liquid from your mash into a bucket or container large enough to hold the enzyme-rich wort.

STUCK MASHES

A stuck mash occurs when the flow of wort from the mash tun stops or slows to a trickle. It typically occurs only when using a large percentage of wheat, rye, or oats. While you can fix a stuck mash, the best approach is to try and avoid one in the first place. Adding half a pound of rice hulls to a mash will help considerably, as will doing a mashout. If you're using a lot of rye or oats, you also can perform a 30-minute rest at around 117°F to break down their sticky beta-glucans.

If you still have a problem with your runoff, try blowing back through the tubing attached to your mash tun. This will remove compacted grain from around the strainer and usually will restart the flow. Then keep the flow on the slow side to prevent compacting the grain bed. If that doesn't work, you may have to scoop out the grain into a bucket or cooler, clean out the mash tun and screen, and start over with some additional rice hulls.

2. Add enough of your sparge water—preferably with the pH adjusted to below 6 (see "Why pH is Important" on page 58)—to the mash so it's loose enough to stir, about 1 gallon for a 5-gallon batch. If you're mashing in a cooler, transfer the grain to your boiling kettle or an appropriate-size pot. Crank up the heat and—stirring continuously—bring it to a boil. Reduce the heat to a simmer and boil the mash for 10 to 30 minutes. (Dark beers need the longer time.) Stir the boiling mash every few minutes to prevent scorching.

3. Turn the heat off. Transfer the reserved wort back to the mash, adding cold water as necessary to hit a temperature of 160–167°F. This will allow the alpha amylase enzymes to convert any starch freed up from the boiling of the decoction. Wait 20 minutes and begin sparging.

Parti-Gyle Mash

This is the way that beer was brewed in the British Isles for several hundred years. A few modern brewers, such as Fuller's, still use this archaic method, which dates back to a time when breweries had huge mash tuns and several smaller kettles. The brewer took the runnings of the first mash and added them to the first kettle, the second runnings to the second kettle, and then the final runnings to yet another kettle. Each kettle had its own hop additions, and each part of the batch fermented separately. The brewer blended these three separate beers together in various percentages to make different beers. One mash could result in an ordinary bitter, a best bitter, a special bitter, a strong ale, and a barleywine.

This type of mashing comes in handy when you're brewing strong beers such as barleywine or imperial stout. It seems so wasteful to cut off the sparging when the gravity of the wort is still coming from the mash at 1.060. You don't want to dilute your imperial stout, but that's still potential tasty beer coming out! You simply continue sparging, collecting the wort until you get the gravity you want or the runnings drop below 1.008/2° Brix on your refractometer. Then treat it like a separate beer, figure out the appropriate amount of hops, and boil it as you normally would. In a recent small-batch barleywine, I collected 5 gallons of wort at 1.120, and then, instead of dumping the grain, I continued to sparge and ended up with 10 gallons of water at 1.045 wort. That's a lot of potential beer! So why doesn't every homebrewer do this? Well, the downside of parti-gyle is that to take advantage of it, you need multiple kettles and burners as well as lots of fermentation space.

A simplified parti-gyle mash that yields two different beers with one mash is a great way to get two beers with only a slight increase in the length of the brew day. If you have two burners and two kettles, it's child's play. Simply stagger the boils so that you can swap out your wort chiller from one kettle to the next at the right time. If you have only one kettle and burner, it'll have to be a long brew day. If you want to give this a try, turn to page 64 for a recipe that yields 5 gallons of imperial stout with a bonus 5 gallons of Irish stout.

To modify this technique, add fresh grain before sparging for your second beer. This allows you to beef up the second runnings or add the color and flavor that your primary mash was lacking. Thus, your first runnings could be an amber-colored barleywine and your second runnings could be a black stout. As long as the malts are in the crystal malt or roasted malt families, they don't need to be converted, and they can be added to the mash without worry.

Parti-Gyle Gravities

This first table shows the gravities of the first third of the wort drawn off (2 gallons of a 6-gallon batch) and the gravity of the rest of the wort (the remaining 4 gallons.)

FULL BATCH OG	FIRST RUNNINGS (⅓)	SECOND RUNNINGS (⅔)
1.050	1.075	1.038
1.060	1.090	1.045
1.070	1.105	1.053
1.080	1.120	1.060

The table below shows the gravities of the first half of the wort drawn off (3 gallons of a 6-gallon batch) and the gravity of the rest of the wort (the remaining 3 gallons.)

FULL BATCH OG	FIRST RUNNINGS (½)	SECOND RUNNINGS (½)
1.050	1.067	1.033
1.060	1.080	1.040
1.070	1.093	1.047
1.080	1.106	1.053

SPARGING

Once you mash, you need to separate the sweet wort you've just created from the spent grains that are no longer needed. Unless you plan to use the no-sparge method, don't forget to *vorlauf,* or recirculate! This is a German brewing term most commercial brewers still use. Basically, it means that when you're finished mashing, you want to recirculate the wort back through the mash until it's fairly clear, with no pieces of malt in it. Then you can run it off to collect your first runnings and proceed to the sparge.

No-Sparge Method

This method is just what it sounds like: At the end of your mash, before you vorlauf, pour the volume of water you'd normally use to sparge right into the mash tun. Give it a stir, do a quick vorlauf, and drain it out. It's not only quick, but it also will eliminate any chance of harsh off-flavors extracting from the mash during sparging. Many homebrewers also note an improvement in malt flavor.

Sounds like a win-win, right? Well, the downside is that this inefficient method involves leaving fermentable sugars behind. To compensate, you typically need to increase your grain bill by about 30 percent. That means almost no commercial brewers skip sparging altogether— 30 percent is a big hit to the bottom line! But on the homebrewing side, the inefficiency is less of an issue. If you're ordering grain in bulk, what's an extra few dollars for a faster and less risky brew day?

PROS ➡ It's speedy, and the increased malt flavor might give you a slight edge in a competition.
CONS ➡ You need a big mash tun, and it's more expensive.

Batch Sparging

This is similar to the no-sparge technique, but the water is added in two or three stages, and you do get some additional sugars rinsed from your grain. First, you drain the wort from the mash, then you add a portion (typically half or a third) of your sparge water and restir the mash. Recirculate the wort until it's clear, and drain the mash again. Repeat until you collect the proper amount of wort.

PROS ➡ It's fairly quick and good for getting all the sugars from the nooks and crannies of a mash— especially with mash tuns that have a single strainer.
CONS ➡ It's tricky to work out the proper amount of water to add in order to get the desired amount in the kettle.

Fly Sparging

The classic version of sparging, this is what commercial breweries do. At home, it works well with mash tuns that have a false bottom covering the entire bottom. The process works like this: Slowly run off your wort and replace the water on top of the mash with sparge water, trying to balance the output from your mash tun to the input of sparge water. Continue until you collect the desired amount of wort.

PROS ➡ It's easy once you calibrate it—you just keep going until you get the volume you need.
CONS ➡ You run the risk of oversparging and extracting tannins. This method also may result in lower extraction when using mash tuns with a single strainer.

WATER ADJUSTMENT

You can't talk about mashing without talking about water. Commercial brewers always know what's in their water. Larger craft brewers sometimes take the water source into consideration when building a new brewery! It makes sense because water is one of the core ingredients in beer.

Homebrewers, on the other hand, tend to fall into two camps: those who know nothing about their water and those who obsessively adjust their water. This could be because most homebrewers start by brewing with malt extract. When you use malt extract, the minerals contained in the water at the malt-extract manufacturer were concentrated with the malt itself. So the brewing water was embedded with the powder or liquid extract.

The nice thing about homebrewing in the twenty-first century is that good, clean water is pretty easy to come by. If you can't filter your own water at home to your satisfaction, the local grocery store will have reverse-osmosis water or filtered water of a comparable quality. Unless you buy reverse-osmosis water or bottled water, you should read on about removing chlorine. Other than that, taking your water chemistry any further is optional.

Removing Chlorine

The single most important thing you should do to your water is remove any chlorine. (If you're using well water, you shouldn't have to worry about chlorine, but check the iron levels). Removing chlorine can be accomplished in a few ways:

➡ **Use a carbon-water filter:** You can buy a carbon-filtering device at your local home-improvement store. The only equipment you'll need are a few feet of old siphon tubing and a garden-hose attachment. Don't run your filter too fast—some chlorine could sneak by. The rule of thumb is 1 gallon of water per minute.

➡ **Boil the water ahead of time:** Bringing water to a boil and turning the heat off is enough to volatilize any chlorine. However, many cities now use chloramines in the water instead of chlorine, which can be more difficult to drive off by boiling. In this case, filtering is a better option.

➡ **Add potassium metabisulfite tablets (Campden tablets):** These tablets are sold by most homebrew suppliers, and one tablet will treat up to 20 gallons of water. Just follow the instructions for the tablets you buy. Most tablets remove chlorine in a few minutes.

Measuring and Adjusting pH

In addition to dechlorinating the water, I like to use a technique that I learned from the Sierra Nevada Brewing Company. They lower all their brewing water to a pH of 5.5, which ensures that no harsh tannins can be extracted. (See "Why pH is Important" on page 66.) The alternative is to adjust the mash pH by using acid malt or adding different minerals to the mash, but this works only up to a point, and you still have to deal with the sparge water's pH separately. The other nice thing about adjusting your brewing water is that you have to do it only once to find out how many milliliters of acid you need—then you can add that amount every time. That means you either can borrow a pH meter from someone or purchase some good pH-indicator strips such as ColorpHast.®

The night before you brew, start with 5 gallons of water and add 2 milliliters of phosphoric or lactic acid (available from your local homebrew supplier). A plastic syringe with milliliter markings works best. Make sure you either stick with the same acid or figure out the amounts for both, because they're different concentrations. Stir well and wait for 5 minutes, then test with your meter or paper strips. If you're above 6.0, then add another milliliter and repeat until your pH falls between 5.5 and 6. Write down the amount of acid needed so you can repeat it in the future without having to take readings. For example, my brewing water requires 3 milliliters of phosphoric acid (10 percent) per 5 gallons to bring the pH to 5.5. The only time you shouldn't use this technique is when brewing beers with very large amounts of dark malts (such as an imperial stout), because the roasted malts are very acidic and could drop the pH too low. Still adjust the sparge water pH, however, to less than 6 to prevent tannin extraction.

Adjusting Minerals

Some brewers add minerals to adjust the pH of their mash. This works to a certain degree and will drop the pH of the mash slightly. But you're also affecting the flavor of the beer by adding so many minerals, and it won't be enough if you have a very high water pH (more than 8). For that reason, I recommend adjusting all your brewing water to the proper pH before the mash as described previously.

You may want to add minerals for other reasons. Yeast cells require 50 parts per million (ppm) of calcium and 10 ppm of magnesium for optimal health, and local water supplies can lack these levels. Even outside yeast health—which certainly affects flavor—minerals affect the taste of the beer by enhancing the malt or the hop character, depending on the mineral. It's a subtle effect, but it allows the brewer to fine-tune the final taste of the beer.

If you're going to adjust the minerals in your water, you need to know the water you're working with. If you buy reverse-osmosis water, you can assume that all of your minerals will be close to zero. If you're using city water, you'll need to request and read a water report. Here's an example of a water report:

Water Report for Asheville, NC

Test	Result
Total dissolved solids (TDS)	57 PPM
Sodium	14 PPM
Calcium	< 1 PPM
Magnesium	< 1 PPM
Total hardness	4 PPM
Sulfate	< 1 PPM
Chloride	7 PPM
Carbonate	< 1 PPM
Bicarbonate	38 PPM
Total alkalinity	31 PPM

WHY pH IS IMPORTANT

Many of the enzymatic and chemical reactions in brewing are affected by pH. For instance, the conversion of starch into sugar during mashing heavily depends on pH. Also, you can avoid the extraction of harsh tannins during sparging with the proper pH. Hop isomerization and beer stability also are affected by pH.

When you brew with water with a pH between 5.5 and 6.0, your mash naturally will fall into the appropriate range of 5.1 to 5.4. The sparge water will never extract tannins from the grain because the pH of the sparge water is adjusted as well.

How much you need to worry about pH depends on your water supply. If you're buying reverse-osmosis or spring water from the store, then you're probably fine without adjusting the pH, but some communities have very alkaline water that will make poor-quality beer if you don't adjust the pH. For example, when I was brewing in New Orleans, the tap water's pH was about 10 to avoid extracting lead from the old pipes. Brewing with that water without any adjustment would have created all kinds of problems.

The important lines in the water report are calcium (Ca), magnesium (Mg), chloride (Cl), and sulfate (SO_4) levels. As you can see in my local water, the calcium is below 1 ppm. That's troubling, because yeast cells require at least 50 ppm for optimal health. Malt does contribute some calcium, but I'd like to see the water itself above 50 ppm. Calcium chloride and calcium sulfate (gypsum) are both forms of calcium and can be added to get the overall calcium levels to acceptable levels. You could use either one, but your choice can affect the flavor of the beer. (If you're making a hop-forward beer, use calcium sulfate. If you're making a malt-forward beer, use calcium chloride.) Magnesium is another mineral necessary for proper yeast health. You generally want 10–30 ppm, and, as you can see, my water contains less than 1 ppm. I would increase that to at least 15 ppm using Epsom salts (magnesium sulfate).

Commercial brewers often talk about the amount of sulfate to chloride as a ratio. A 3:1 ratio is considered optimal for pale ales and IPAs, a 1:1 ratio is best for balanced beers, and a 1:3 ratio is best for malty beers. The Epsom salts you use to bring up the magnesium levels also contain sulfate, so consider that as part of your numbers.

If all this sounds complicated, don't worry! Nobody figures out their mineral additions on paper. The Internet offers free water calculators. I often use the EZ Water Calculator (ezwatercalculator.com). It allows you to adjust amounts up and down until you get the water profile you want. Since my water is almost as devoid of minerals as reverse-osmosis or distilled water, I have a blank slate to work with. At right are two water profiles that I use, which you can use as well if you start with similarly neutral water.

Water Profiles

Type	Water character	Typical beers	Calcium	Magnesium	Sulfate	Chloride
Burton	Hard, good for accentuating hops, too much sulfate for most beers	Bitters, IPAs	352	24	820	16
Dublin	High levels of bicarbonate, good for dark beers	Stout	118	4	54	19
London	Balanced	Bitters	52	32	32	34
Munich	Balanced	Amber to dark lagers	109	21	79	36
Pilsen	Almost no minerals, very soft	Pilsners, dark lagers	10	3	0	4
Typical soft water	Needs mineral additions for yeast health		0	0	6	3
Ideal water for hoppy beers	High sulfate-to-chloride ratio enhances hops	Pale/IPA	60-80	15	150-175	50-60
Ideal water for malty beers	High chloride-to-sulfate ratio enhances malt	Malt-forward beers	60-80	10-15	50-75	100-150

Note: While some of these water profiles are taken from the cities' data, most brewers don't use the water as-is. Most either treat their water, blend water from multiple sources, or do both, depending on the beer.

Hop-Forward Water Profile

Goal: More than 50 ppm calcium and more than 10 ppm magnesium, which is good for yeast health, and a 3:1 sulfate-to-chloride ratio, which will make the hops pop.

Target numbers: 65 ppm calcium, 18 ppm magnesium, 164 ppm sulfate, and 58 ppm chloride

Add to mash: 2 grams gypsum, 1.5 grams calcium chloride, 2.5 grams Epsom salts

Add to kettle: 2.9 grams gypsum, 2.1 grams calcium chloride, 3.6 grams Epsom salts

Malt-Forward Water Profile

Goal: More than 50 ppm calcium, just more than 10 ppm magnesium, which is good for yeast health, and around a 1:2 sulfate-to-chloride ratio, which will enhance the malt character in bocks, brown ales, and similar beers.

Target numbers: 62 ppm calcium, 11 ppm magnesium, 50 ppm sulfate, and 112 ppm chloride.

Add to mash: 3 grams calcium chloride, 1.5 grams Epsom salts

Add to kettle: 4.3 grams calcium chloride, 2.1 grams Epsom salts

UMLAUT DUNKEL

This is one of my favorite beers to have on tap at home. Its alcohol content is so low that I don't feel guilty about breaking out the half-liter mug, but this beer still has plenty of malt complexity. German and Czech brewers would decoction mash this one. If you have the urge to follow their lead and be traditional, see page 50 for more details. Otherwise, a simple infusion mash will work fine thanks to the melanoidin-rich malts in the grain bill.

YOU NEED
basic brewing equipment (page 3)

8½ gallons filtered brewing water (page 12)

6 pounds German dark Munich malt (60%)

3 pounds German Vienna malt (30%)

¾ pound German Caramunich III malt (7.5%)

¼ pound German Carafa Spezial II (2.5%)

4 alpha acid units German Hallertau hops, or any other German hops, at 60 minutes (14.5 IBU)

1 Whirlfloc tablet

3 vials or packages German Lager WLP838/ Bavarian Lager WY2206 yeast (or a 2-liter starter made from 1 pack, page 110)

3¾ ounces dextrose/corn sugar (optional, use only for bottling)

TARGETS
Yield: 5 gallons
OG: 1.048–1.050
FG: 1.012
IBU: 14.5

1. Mix the malt with 3½ gallons of water at 168°F or the appropriate temperature to mash at 153°F. Mash for 60 minutes.

2. Recirculate the wort until it's fairly clear. Run off the wort into the kettle.

3. Sparge with 5 more gallons of water at 165°F. Run off the wort into the kettle.

4. Bring the wort to a boil. Boil it for 30 minutes. Add the hops and continue to boil for 60 minutes. Add the Whirlfloc tablet at 30 minutes. Put your wort chiller into the wort at least 15 minutes before the end of the boil.

5. When the boil finishes, cover the pot with a lid or a new trash bag and chill to 48°F if possible. If you can't chill that low, get it as cool as possible, siphon the wort into your sanitized fermentor, and cool it in your lagering fridge overnight. When the wort hits 48°F, pitch three packs of liquid yeast or a 2-liter starter.

Note: When brewing lagers, it's very important to chill to fermentation temperatures before oxygenating and pitching the yeast.

6. Ferment at 48°F for two days, then let the temperature rise to 50°F for another three days. Raise the temperature to 55°F for seven days, then raise it again to 65–68°F for the last three days. Crash the temperature down to 32°F for one week.

7. Keg or bottle the beer. (If you're bottling, I recommend 3¾ ounces of dextrose/corn sugar for this beer.)

BLACK

I'm always changing the name of this beer. I've called it Black Sabbath, Black Sunday, Black Magus, and dozens of other things. The name may change, but the recipe is dialed in for the perfect black ale to have on tap at my house. What kind of beer is it? It's comparable to a foreign-export stout. But it sounds weird to call it that when it's neither foreign nor exported, so I just refer to it as a black ale.

YOU NEED

basic brewing equipment (page 1)

10 gallons filtered brewing water (page 12)

11 pounds North American 2-row malt (73%)

1½ pounds British black malt (10%)

1 pound dark Munich malt (6.7%)

1 pound flaked barley (6.7%)

½ pound U.S. crystal 80°L (3.3%)

13 alpha acid units Columbus hops at 60 minutes (36 IBU)

8 alpha acid units Columbus hops at 30 minutes (17 IBU)

1 Whirlfloc tablet

2 vials or packages California Ale WLP001/ American Ale WY1056 yeast (or a 2-liter starter made from 1 pack; page 110)

3¾ ounces dextrose/corn sugar (optional, use only for bottling)

TARGETS
Yield: 5 gallons

OG: 1.063–1.068

FG: 1.014

IBU: 66

1. Mix the malt with 5 gallons of water at 165°F or the appropriate temperature to mash at 150°F. Mash for 60 minutes.

2. Recirculate the wort until it's fairly clear. Run off the wort into the kettle.

3. Sparge with 5 more gallons of water at 165°F. Run off the wort into the kettle.

4. Bring the wort to a boil. Boil it for 15 minutes. Add the hops and continue to boil for 60 minutes. Add the Whirlfloc tablet and the other hop addition at 30 minutes. Put your wort chiller into the wort at least 15 minutes before the end of the boil.

5. When the boil finishes, cover the pot with a lid or a new trash bag and chill to 65°F. Siphon the wort into your sanitized fermentor and pitch two packs of liquid yeast or a 2-liter starter.

6. Ferment at 65°F for one week, then let it warm up to 68–70°F for the second week.

7. Keg or bottle the beer. (If you're bottling, I recommend 3¾ ounces of dextrose/corn sugar for this beer.)

IMPERIAL STOUT/ IRISH STOUT (PARTI-GYLE)

This beer is based on a recipe from the Courage Brewery that dates back to 1914. It uses only brown malt and black malt as specialty malts. Because the original recipe calls for a huge amount of Fuggle hops, I substituted a hop with a higher alpha-acid percentage so I'd have less hop residue in the kettle. This beer is so intensely flavored that the hop variety is pretty much irrelevant anyway. The imperial stout is inky black with an intense bittersweet-chocolate flavor. At more than 10% ABV, it'll be good for a decade, so bottle it and sample the bottles over the years. As a bonus, you get a free Irish Stout from the second runnings. It's a low-gravity stout similar to Guinness that's the very definition of a session beer: easy drinking yet complex.

YOU NEED
basic brewing equipment (page 3)

18 gallons filtered brewing water (page 12)

20 pounds British pale-ale malt (76.9%)

4 pounds brown malt (15.4%)

2 pounds British black malt (7.7%)

30 alpha acid units Columbus hops at 75 minutes for the imperial stout (131 IBU, calculated)

1 Whirlfloc tablet

8 alpha acid units Target hops or any British hops at 75 minutes for the Irish Stout (35 IBU)

4 vials or packages Irish Ale WLP004/WY1084 yeast (or a 1-gallon starter made from 1–2 packs or pitch on the yeast cake from a previous batch; page 110) plus an additional 1 vial or package for the Irish Stout made from the second runnings

3¾ ounces dextrose/corn sugar (optional, use only for bottling) or 7½ ounces if bottling both batches

Note: Dry yeast also is ideal for a beer this big. Safale US-05 is a good choice, and two 11.5-gram packs provide enough cells.

TARGETS (IMPERIAL STOUT)
Yield: 5 gallons

OG: 1.097–1.102

FG: 1.020–1.025

IBU: 131 (calculated)

TARGETS (IRISH STOUT)
Yield: 5 gallons

OG: 1.038–1.040

FG: 1.008–1.010

IBU: 35

1. Mix the malt with 8 gallons of water at 175°F or the appropriate temperature to mash at 150°F. Because of the huge amount of grain in this recipe, you need to heat the mash water higher than normal. Mash for 60 minutes.

2. Recirculate the wort until it's fairly clear. Run off the wort into the kettle.

3. Sparge with enough water at 165°F to collect 6½ gallons of wort. You should have a gravity of around

1.086. Boil the wort, adding the Columbus hops at the beginning of the boil. Boil for 75 minutes, adding the Whirlfloc tablet with 30 minutes left in the boil. Put your wort chiller into the wort at least 15 minutes before the end of the boil. You should end up with 5½ gallons of OG 1.100 + wort.

4. Meanwhile, collect the Irish Stout wort. Add 3 gallons of 165°F sparge water into the mash and stir well. Recirculate until fairly clear, then start to collect the runoff. Continue to sparge until you collect 6½ gallons of wort. You should have a gravity of about 1.030–1.032. When you're done boiling the imperial stout, you'll boil this up, adding the Target or Northern Brewer hops at the beginning of the 75-minute boil. You'll end up with an OG of 1.038–1.040. Proceed as with a typical brew, adding one vial or package of yeast. Ferment and package just like the imperial stout. You can start drinking the Irish Stout right away.

5. When the imperial stout finishes boiling, cover the pot with a lid or a new trash bag and chill to 65°F. Siphon the wort into your sanitized fermentor and pitch four packs of liquid yeast, two packs of dry yeast, a 1-gallon starter, or the yeast cake from a previous beer.

6. Ferment at 65°F for one week, then let it warm up to 68–70°F for the second week.

7. Keg or bottle the beer. (If you're bottling, I recommend 3¾ ounces of dextrose/corn sugar for this beer.) Age the imperial stout for at least three months before sampling.

HOPS and HOPPING

HOPS

WHEN I STARTED BREWING in the late 1980s, hops came in the form of hops extract already added to the malt extract. If you could find hops at all, they usually were sitting on a shelf in a plastic bag and often several years old! Fast-forward to today: hop processors package hops in nitrogen-purged foil bags, and homebrew shops keep them refrigerated or frozen until you buy them. The varieties are mind-boggling. Not only do we have dozens of unique North American hops, but hops also come from exotic locales such as New Zealand, Slovenia, and Patagonia. With interesting new varieties coming out every year, this is definitely the golden age for hoppy-beer lovers.

HOP PROPERTIES

Hops look like little pinecones. They grow on the bine *Humulus lupulus*, which can measure up to 20 feet tall. If you break open a ripe hop cone, you'll see a yellow-orange powder in the center; this is the good stuff. It contains the resin glands of the hop, with both the bitter acids and aromatic oils that brewers desire.

Alpha Acids and Bitterness

The amount of alpha acids present in the lupulin glands determines the bittering ability of a hop. Originally, most hops had a low bittering potential, with only 2–3 percent alpha acid (AA). However, commercial breeding has increased the upper limit to close to 20 percent. Since you need 9½ ounces of a 2 percent AA hop to equal the bitterness of 1 ounce of a 19 percent AA hop, you can see why commercial brewers like high-alpha hops for bittering!

The AAs aren't bitter until they isomerize (change on the molecular level) in the wort during the boil. The longer the hops boil, the greater the amount of isomerized AA. No matter how long you boil, the best you can achieve is around 25–30 percent of the total AA being isomerized—and probably less than that in high-gravity wort. This percentage is called "utilization" and depends on the length of the boil and the gravity of the wort. The graph below illustrates how to determine the utilization of your hops.

The alpha acid percentage combined with the utilization creates bitterness in the beer, typically measured in international bitterness units (IBUs). Formulas and calculations can determine your utilization, but not many brewers these days do their bitterness calculations by hand. In addition to the many variables, it's a no-brainer to let inexpensive or free brewing programs do some of the math for you. (See page 191 for my recommendations on brewing calculators.)

The recipes in this book provide the bitterness calculations (IBUs). Additionally, any hops that contribute bitterness appear in alpha acid units rather than in ounces. An alpha acid unit is simply the alpha acid percentage of a hop variety multiplied by the number of ounces of the hop used in the recipe. For example, 2 ounces of Cascade hops with 6 percent alpha acid equals 12 alpha acid units. It's my preferred way to express the amount of bittering hops needed in recipes since hops can vary wildly in alpha acid percentage from crop to crop and year to year.

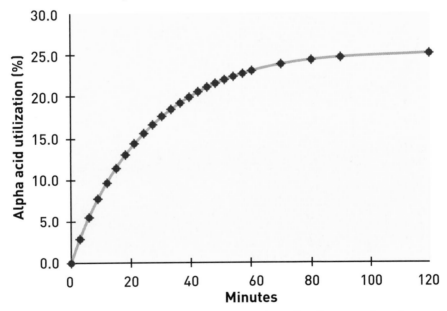

Alpha Acid Utilization vs. Time for 1.050 OG

Graph data courtesy of Glenn Tinseth (http://realbeer.com/hops/research.html).

Note: The percentage of beta acids sometimes is listed for hops, but it doesn't mean much to brewers. Beta acids don't isomerize in the wort to create bitterness in the same way alpha acids do, but they can oxidize slowly over time in the finished beer and add bitterness.

The brewer intent on making the bitterest beer possible should know that increasing bitterness is more complicated than just throwing in more hops. There's a level of saturation above which the wort can't absorb any more bitterness (often touted as 100 IBU, though it's likely a little above or below that, depending on the gravity of the wort). The most important thing to keep in mind is that there's a big difference between calculated and measured IBUs, even with brewing software. I've brewed plenty of beers calculated at more than 100 IBUs, but the IBU in the finished beer was closer to 80 or 90 when tested. In practice, if you have a malt base beefy enough to handle a heavy hop charge, then it's impossible to overbitter it if you love hops. Any all-malt beer above 1.070 should be able to handle whatever you throw at it.

Also keep in mind that it's not only hop aroma that fades. Hop bitterness drops off quickly in finished beer. If you're formulating the bitterness for an imperial stout that will age for a year, double your bittering hops.

ALPHA ACIDS AND COHUMULONE

The alpha acids we're discussing are primarily broken down into three main categories: humulone, adhumulone, and cohumulone. Cohumulones give a harsher bitterness, so brewers tend to look for hops that have a lower percentage of cohumulone compared to the other acids. But is this true or just a brewing myth?

It all started with an experiment in 1972, when researchers sampled beers brewed with isolated acids. But the different alpha acids isomerize differently, so the beers had wildly different levels of isomerized alpha acids and, therefore, bitterness. Cohumulone had the highest isomerization levels and was thus the bitterest. Many brewers misinterpreted the article to mean that cohumulone was harsh and that hops with low cohumulone levels provided better-quality bitterness.

I put this theory to the test with a blind tasting among Beer Judge Certification Program (BJCP)–certified judges and a lab analysis to see whether there was a notable difference in isomerized alpha acids between the two. I brewed two normal pale ales with no mineral additions. For one beer, I used Cascade for bittering, which is high in cohumulone, at 33–37 percent. For the other I used Simcoe®, which is very low in cohumulone, at 17–22 percent. I used only bittering hops, so perceived bitterness from later hop additions wouldn't affect the results. I used an appropriate amount of hops to add 40 IBUs to each beer, so both had the same calculated level of bitterness.

The blind tasting confirmed earlier findings: The judges perceived Cascade-bittered beer as having a stronger bitterness than the Simcoe-bittered beer. But that was only part of the study. The lab analysis showed that the beer brewed with Simcoe had 33 IBUs, while the beer brewed with Cascade had 42 IBUs, a whopping 27 percent more. With that much difference in actual IBUs, it's not surprising that tasters thought the bitterness from low-cohumulone hops was smoother. It was less bitter. Before I make any blanket statements, this experiment needs more data points. But it does seem that hops with lower cohumulone levels contribute fewer isomerized alpha acids than a high-cohumulone hop with the same alpha acid percentage. If this is true, then working cohumulone levels into IBU calculators would make them far more accurate.

Common Bittering Hops

Any hop can be used as a bittering hop. Some of the classic examples of English and German beers are bittered with low-alpha hops. For financial reasons, however, adding a higher-alpha hop for bittering is usually preferred. It costs a brewery less for the same level of bitterness, and less hop sludge in the kettle means more beer going to the fermentor. There's a saying that "alpha is alpha," which means that hop character doesn't survive an hour-plus boil, so you might as well use the highest-alpha hop that you can for bittering. However, the cohumulone experiment (page 70) proves that this isn't necessarily the case. Some hop character definitely can survive a long boil. Thus, the wrong choice of bittering hop can come through in your brew.

Luckily, higher-alpha hops are great for bittering a wide range of beers while adding a complexity that can blend well with later hop additions. Some varieties such as Horizon and Magnum are neutral in flavor and can be used in almost any beer style for a bittering addition. Others, such as German Merkur, have a traditional German flavor and are a better choice for authentic lagers. The British hop Target is a high-alpha hop that still tastes very English and is a great choice for British ales. Use the table below as a starting point, but experiment to find what's right for your beers.

Common Bittering Hops

Hop	Alpha Acid Range	Reason Used	Typical Beer Styles
Warrior, Nugget, Horizon, Magnum	15–18%	These hops provide a clean bitterness that stays from the way for aroma hops. The high alpha acid percentage means a big bang for your buck.	Many American ales, especially pale ale, IPA, and other hoppy beers
Chinook	11–13%	This hop imparts a crisp, in-your-face bitterness, stronger than most other high-alpha hops. This can be a real benefit for IPAs and other aggressive beers.	American IPA, imperial IPA, American black ale, etc.
Summit™	16–18%	With one of the highest AA percentages available, this hop can be oniony or garlicky; definitely dank.	Use only in aggressive West Coast–style pale ale or IPA.
Columbus	12–16%	Can be somewhat dank but more refined than Summit, with minty, perfumy notes.	Good choice for any West Coast–style IPA, red ale, etc.
Centennial	8.5–11.5%	Refined citrus and fruit, fruit-flavored cereal.	A classic hop for pale ale and IPA
Target	9.5–12.5%	Good British flavor with higher than typical AA percentage.	Any UK ale
German Merkur, Hercules	12–17%	Huge increase in AA percentage compared to Hallertau or Saaz, while retaining an authentic German character.	Any traditional German ale or lager

Oils and Aroma

Even though alpha acids are one of a brewer's primary concerns, other components of the hops also are very important, depending on the beer style. Think of hops as you would chiles: They're all hot to differing degrees, but their flavor often adds as much to a dish as their heat. The delicate essential oils are perhaps the most important because the aroma and taste of hops come from them. Almost all of these oils evaporate or are otherwise lost during a full 60-minute boil, so it's essential to add hops later in the boil if you want to retain some of their aroma and taste.

If you add hops during the last 10 to 30 minutes of the boil, you'll lose most of the aromatic components but will retain some of the hop flavor. That's why these additions often are called "flavor-hop additions."

Hops added during the last few minutes of the boil are finishing or aroma hops. Even a short time in hot wort will volatilize a lot of the aromatics. This isn't

Common Aroma Hops

Hop	Strength of Aroma	Type of Aroma	Typical Uses	Alpha Acid Range
Cascade	Medium	Classic grapefruit	Many American ales, especially pale ales	7–9%
Centennial	Medium	Fruity, floral	More refined than Cascade but used in similar beers	8.5–11.5%
Simcoe	High	Pine, cat pee, marijuana	Essential in many West Coast IPAs	12–14%
Amarillo	Medium-high	Tropical fruit, citrus, apricot, peach	Any beer that needs a juicy hop aroma and taste, popular in blends	8.5–11%
Citra®	High	Mango, papaya, citrus	Very pungent and popular in IPAs by itself, often blended with other hops for pale ales and other beers	11–13%
Palisade®	Medium	Fruity, grape juice	Wheat beers, saisons, Belgian blondes	5.5–9.5%
Columbus/CTZ	High	Herbal, mint, citrus, dank	Aggressively hopped West Coast styles	14–16%
East Kent Goldings/Fuggle	Low	Fruity, leather, tobacco	Any British beer style	4.5–5.5%
Galaxy (Australia)	High	Passion fruit, citrus, dank	Aggressively hopped West Coast styles	13.5–15%
Sterling	Low-medium	Floral	Pilsners, helles, American lagers	6–9%

Note: For more complete descriptions of these and other hops, see page 77.

necessarily a bad thing. German brewers don't care for those types of aromatics and usually will add their final hop addition 10 to 15 minutes before the end of the boil. If you're trying to keep as much of the aroma as possible in the beer, consider dry hopping (page 85).

Note: Brewers making IPAs and other hop-centric beers are especially interested in the oil content of hops. Most hop processors don't list hop oil levels on their packaging, but their websites often contain this information. It's very difficult to list specific oil percentages for individual hops because they vary widely from year to year and even farm to farm. For example, from 1975 to 1985 the total oil content in Cascade hops varied from 0.28 to 1.79 milliliters per 100 grams of hops, and the most prevalent oil, myrcene, varied from 46 to 82 percent. Unless you can get an accurate analysis of a particular crop, you'll have to make an educated guess.

Common Aroma Hops

While "aroma hops" can be a nickname for any hops added at the end of the boil, hops valued for their aromatic properties also are called "aroma hops." Their alpha acid levels are of lesser concern for most brewers because generally they're added at the end of the boil and don't impart much bitterness. But that doesn't mean high-alpha hops aren't used frequently for aroma. A few of the most popular hop varieties for dry hopping intense IPAs also have some of the highest alpha acid levels. It makes sense: High levels of alpha acid often correspond to high levels of aromatic oils.

It may be tough to find or afford a particular hop from year to year as it waxes or wanes in popularity. Keep in mind that, as a homebrewer, you can take risks that some of the big breweries can't. Popular hops may be the safe bet, but you easily can brew an IPA with an experimental new hop or an overlooked older hop and get great results. Why brew what everyone else is brewing? That said, certain aroma hops have stayed popular for years—or even centuries, in the case of the noble hops.

HOP FORMS

Whole Cone

Whole-cone hops, also known as leaf hops, are dried, unprocessed hop cones. Most craft brewers use pellets, but a few exclusively use whole-hop cones. Brewers who want to use whole hops often have a smaller selection of varieties to choose from compared to pellets. Whole hops also are more susceptible to degradation, so cold storage is essential. I wouldn't purchase whole hops after August or September. At that point, wait for the new crop to come out, which usually happens in late October. Once you open a vacuum-sealed pack of whole-cone hops, store them in the freezer in double-layered freezer bags, with as much air squeezed out as possible. Use them within two to three months.

The whole-cone hops you buy at homebrew stores or online are dried by the processor before they're packed and shipped. In other words, you'll always be using dried hops. Wet hops are undried hops taken right off the bine. They must be used within hours of harvest. Unless you grow your own hops or have a local hop grower, you probably won't have access to wet hops. You'll see wet-hop beers in the fall (hop harvest season in America) from many craft brewers, however, and it's a good idea to sample some of these to see how the wet hops taste compared to typical dried hops. Most people describe them as grassier, earthier, and lacking the fresh citrus aroma of dried hops. Nationwide, Sierra Nevada may be your best option. (They have a variety of harvest and wet-hop ales.) Rogue Ales also produces a few wet-hop beers every year, as do many smaller regional brewers.

Note: Whole hops sometimes are called "fresh hops" mistakenly—perhaps because a well-known company that supplies whole hops to homebrewers goes by that name. The term "fresh hop" has no agreed-upon definition, unlike wet hop.

PROS ➡ Some brewers feel that pelletizing hops damages the delicate oils and that whole hops have a better aroma. Whole hops also can form a filter bed in the bottom of the kettle, which can help remove break material.

CONS ➡ Unless stored in the freezer in appropriate packaging, whole hops degrade quickly. They lose bitterness and hop aroma potential and can take on a cheesy aroma much faster than their pellet counterparts.

Pellet

Pellet hops are the most common form of hops. They're made when hop processors remove some of the leafy material from whole-cone hops, grind the remaining cone into a powder, and form it into pellets. The most common pellet is the T-90 pellet. The name T-90 indicates that 90 percent of the original hop is present in the pellet. Most of the hops you buy at your homebrew store are this variety. A less common pellet type is T-45, which contains only 45 percent of the original hop matter. This type of hop concentrates the oils and the bitter lupulin, therefore requiring fewer total ounces for the same effect. I have seen T-45 pellets used only once in my career: When Czech Saaz hops were down to 1.5 percent alpha acid, a local brewer of traditional pilsners switched to T-45 to limit the amount of hop residue in the kettle.

Once you open a vacuum-sealed pack, store pellet hops in a glass Mason jar in the freezer. If you minimize headspace—or even better, purge the headspace with CO_2—they'll keep fresh for about a year. Some hops are more susceptible to degradation than others, which is expressed as the Hop Stability Index (HSI). The HSI tells you how much of the alpha acids will degrade in six months at 60°F, which can range from 10–50 percent, depending on the variety. At freezing temperatures the loss drops significantly, however, which is why cold storage is so essential.

PROS ➡ Pellets stored properly (vacuum sealed) can last for years with minimal degradation. They take up much less room than whole hops and are available in a huge selection of varieties.

CONS ➡ Unless you put them in a fine-mesh bag, pellets can make it difficult to separate the hop residue from the wort in the kettle. Also, using pellets isn't as much fun as throwing armloads of dried flowers into the kettle.

Extract

Hops extracts are used primarily by larger breweries. Until recently, craft brewers looked down on them. But the rise of double IPAs has made small brewers think twice about using hops extracts, because the benefit of not having to separate out a huge amount of hop residue from the wort in the kettle means more beer going to the fermentor, and more beer in the fermentor translates into more pints of beer per batch, which makes everyone happy. Homebrewers now have access to some extracts in homebrew-size containers. Usually they're packaged in plastic syringes that allow a brewer to add a predetermined amount of bitterness into the kettle. The syringes don't deliver the advertised bitterness, so you'll need to add at least 25 percent more to get the desired bitterness. Hops extracts also can fix under-hopped beer. As long as the extract is isomerized, you can add it to taste to finished beer.

Note: Hop-oil extract, which gives hop aroma and taste but no bitterness to beer, isn't recommended. Almost all varieties have a fake, perfumy character.

PROS ➡ Less vegetative material in the kettle means more beer in the fermentor.

CONS ➡ There's little to no choice in variety of hops, so these extracts should be used for bittering only.

Hop Hash

A new product on the market, hop "hash" is the residue scraped off the hop-pelletizing machine. It used to be thrown away but now is in very high demand and can be difficult to source. Cryo Hops®, a similar product, is more of a sifted lupulin gland powder but should be treated in the same way. It can be used at the end of the boil or as a dry hop, although it can be hard to get into solution. When dry hashing, mixing the hash with 1 cup of vodka will help a lot.

PROS ➡ The aroma is intense and there's less beer lost due to less vegetative material.

CONS ➡ It's rare and difficult to work with as a dry hop; you may need to make a vodka solution.

HOP VARIETIES

Some hop varieties date back more than 100 years, whereas other varieties came straight from breeding programs in the last decade. On the following pages you'll find a breakdown of some of the most commonly available hop varieties. There will always be new varieties, but most of these hops are well established and will be around for decades.

American Hops

Amarillo (6–9% alpha acid): Fantastic peach/apricot and tropical-fruit aromas make this a go-to hop for big, hoppy red ales or pale ales. It's a hybrid of Cascade, so it plays well with that hop. For huge hop character, try the modern classic of 50 percent Amarillo and 50 percent Simcoe for dry hopping.

Apollo (15–19% alpha acid): Similar to Summit, this is a bittering monster, but the onion/garlic notes may turn off some brewers when it comes to late additions.

Azacca® (14–16% alpha acid): A dwarf hop with an intense tropical fruit and citrus character, this is a great dry hop by itself or as part of a blend.

Calypso (12–14% alpha acid): This hop has a unique aroma of pear, apple, and stone fruit. Nice in saisons and English IPAs.

Cascade (6–8% alpha acid): When the Sierra Nevada and Anchor breweries opened, the selection of hop varieties was limited. Both breweries started using this little-known varietal, and it's now recognized as the classic hop for American pale ales and IPAs. (Sierra Nevada's Pale Ale still uses mostly Cascade for aroma.) Generally described as having a grapefruity and citrusy aroma and taste, it can have quite a bit of variation depending on where it grew. Cascade is great for dry hopping, but it's more subdued than many of the newer varieties. One of the great things about Cascade is that, even when other popular aroma hops become scarce, there's generally enough Cascade to go around.

Centennial (8–11% alpha acid): Often described as a super Cascade, Centennial is a very different hop, with a more refined fruity aroma and substantial floral notes. I've brewed pilsners with 100 percent Centennial hops that were wonderful. It's also a great hop for IPAs (Bell's Two Hearted Ale uses 100 percent Centennial.) You can use it when dry hopping, but it's subtler than intense hops such as Citra, Simcoe, and Amarillo. Blend it with a couple of those to create a more balanced but still hoppy profile.

Chinook (11–13% alpha acid): This old-school high-alpha hop variety was once considered too dank and piney when dank and piney weren't desirable. Now that many beer drinkers prefer intense hops, Chinook is quite popular. I love the resiny taste that comes through even when using it as a bittering hop. It's perfect for many dark ales and a great secret weapon in a dry-hop blend.

Citra (10–13% alpha acid): This hop is a personal favorite and very popular with many craft brewers. It oozes mango, papaya, and other tropical-fruit aromas and tastes. It's hugely popular with commercial brewers. In a short time, it's become indispensable in IPAs and imperial IPAs. It's also great for dry hopping, because it comes through strong. It can be overpowering and cloying by itself, but it plays well with other American hops.

Cluster (6–8% alpha acid): Never embraced by homebrewers, Cluster deserves a mention. For many decades, it was *the* hop used in America by almost all the large breweries. It has the taste of vintage American lager and is worthy of experimentation in your light lagers. Not recommended for dry hopping.

Comet (9.5–12.5% alpha acid): This love-it-or-hate-it hop has heavy resin and grapefruit character with a distinct wild American hop aroma.

Columbus (11–18% alpha acid): Generally called Columbus, this hop also grows under the names Tomahawk and Zeus, so brewers sometimes refer to it as CTZ. No matter what name it goes by, it's the same hop grown by different companies. Perhaps because so many places grow it, it tends to have a lot of variation from year to year—and even farmer to farmer. In general, it has a fruity, minty, almost perfume-like intensity that you can smell from across the room. Many commercial breweries use it for its high alphas and aroma. Many also use it for dry hopping but usually blended with other hops.

El Dorado (14–17% alpha acid): Another heavy hitter with high alpha acid that's a little more delicate with melon and mango aromas. Not an IPA hop. Use this in blond ales, wheat beers, and unique Belgian blondes.

Equinox®/Ekuanot™ (14–15% alpha acid): A new hop that boasts lemon or lime peel combined with an unusual green bell pepper note. It's having a bit of a naming problem at the moment, so time will tell its final name.

Falconer's Flight® (9.5–12% alpha acid): One of a few new proprietary hop blends, this one focuses on floral, citrus, and tropical fruit.

Falconer's Flight 7 C's® (9–10.5% alpha acid): This proprietary hop blend focuses on earthy, citrusy, and spicy notes.

Horizon (12–16% alpha acid): A clean-tasting hop that's great for bittering additions due to its high alpha acid percentage, Horizon is fairly neutral in taste, making it a hop you can use for most styles. It isn't used much for aroma or dry hopping, but it does have good levels of oils.

Mosaic® (11.5–13.5% alpha acid): A very popular new aroma hop with big berry and ripe mango notes that blends amazingly well with other hops or shines on its own. If you're growing bored of the usual citrus/pine hops, give Mosaic a try in your next IPA.

Northern Brewer (7–9% alpha acid): This hop probably would have disappeared long ago if not for Anchor Brewing's Steam Beer, which uses it exclusively. It has a distinct woody, earthy aroma and taste, which makes it a nice hop to blend with citrusy or piney hops. I wouldn't dry hop with it on its own, but it could work in a blend with some Amarillo and Centennial.

Sorachi Ace (10–16% alpha acid): Originally cultivated in Japan, this hop now grows in America. Although marketed as being very lemony, it also can impart an intense dill character. Some companies now use the words "herbal" or "dill" as descriptors for Sorachi Ace. Use a small amount and blend it with other hops unless you're working on something like a cucumber saison.

Simcoe (12–14% alpha acid): This hop made it cool to like dank, funky hops. A lot of breweries even brew 100 percent Simcoe beers to showcase its intense aroma and taste. Notorious for variations between crops, some Simcoe harvests have a distinct cat-pee smell, while others have glorious pine and citrus notes. Still others have hints of onion and garlic. If you come across a batch of Simcoe that you love, splurge on a pound or more. Simcoe is almost essential in modern West Coast IPAs and imperial IPAs. If you like huge hop aroma, then try dry hopping with a few ounces of straight Simcoe.

WHAT IS DANK?

Originally a term for high-grade marijuana, "dank" has been adopted by brewers to denote an extremely pungent, oily hop variety. These hops usually have strong pine, herbal, and tropical-fruit characteristics, though some veer into onion and garlic territory. They are de rigueur in West Coast IPAs, imperial IPAs, and double red ales. Thus, many of them have become expensive (and rare) for homebrewers and commercial brewers alike.

The dankest hop varieties usually are high-Alpha varieties with high oil levels. Citra, Simcoe, Summit, and Columbus are the usual suspects, but brewers often finesse the hop aroma by combining a dank hop with a more refined variety such as Cascade, Centennial, or Amarillo.

Summit (17–19% alpha acid): This hop is about as bitter as a hop can get, so it makes an economical bittering hop. Some brewers like its pungent character, but it's too much for me—although in a dry-hop blend it can add some complexity.

Zythos® (10–12.5% alpha acid): This proprietary hop blend aims for a citrus and tropical-fruit character.

British Hops

British hops are all about subtle differences in flavor. Unlike Citra's big tropical fruit and Simcoe's dank citrus and pine, there's not as much over-the-top aroma to coax from these British hops, but it's still worthwhile to experiment and go down the path of creating some classic British Ales. See page 84 for a recipe that goes overboard with British hops.

Challenger (6–9% alpha acid): This is a popular hop among commercial British brewers for both bittering and aroma. It has a light, fruity scent that's great in bitters and British IPAs. It has a minimal dry-hop contribution.

East Kent Goldings, a.k.a. EKG (5–5.5% alpha acid): This is the most traditional hop for classic British ales, with more of a fruity, berrylike aroma and taste compared to Fuggle, especially in the pelletized versions. This character comes through in dry hopping, though it's not as strong as American hops. Occasionally you can find a variety called Whitbread Goldings, which isn't a true Goldings and isn't as refined as EKG. In addition, you'll often find Styrian Goldings, which isn't a true Goldings either but is actually a Fuggle from Slovenia. Belgian brewers love Styrian Goldings.

Fuggle (4–4.5% alpha acid): A classic British ale hop that has a woody, fruity (dried apricot) aroma and taste with hints of pipe tobacco. Fuggle is a personal favorite in any traditional British bitter or pale ale. It has a minimal dry-hop contribution.

Northdown (7–9% alpha acid): This hop can be used for both bittering and aroma in the kettle. It derives from Northern Brewer and has a hint of the minty, woody character that Northern Brewer is known for. It's especially good in darker ales. Not recommended for dry hopping.

Target (9–12% alpha acid): Target is ideal for bittering any British ale. It has a firm bitterness and blends perfectly with Fuggle or Goldings later in the boil. Not recommended for dry hopping.

German and Czech Hops

Four hops have the designation of noble hops: Hallertau, Saaz, Spalt, and Tettnang. These are the traditional hops used in lager brewing. They're characterized by a low alpha acid percentage and low levels of cohumulone. Noble hops are admired for their smooth bitterness and delicate aromas. They should be your go-to hops for any classic European lager. Some exciting new hybrids crossbreed traditional noble hops with American Cascade hops with interesting results. I usually don't recommend dry hopping with noble hops, but these new hybrids can work well with subtle dry hopping.

Hallertau (2.5–4.5% alpha acid): This classic German lager hop has an aroma of fresh straw and sometimes a hint of chamomile. The name Hallertau refers to the growing region, so you may see names such as Hallertauer Mittelfrau, Hallertauer Hersbrucker, and even Hallertauer Hallertau. Some subtle differences exist among the varieties, but they all can be substituted for one another in a recipe. American Hallertau, however, isn't an appropriate substitute for the real (German) deal!

Hallertau Blanc (9–12% alpha acid): One of the most popular of the new German hybrids, this one pairs a hefty bittering potential with the aroma of white grapes, citrus, and lemongrass.

Huell Melon (6.5–7.5% alpha acid): Another of the new German hybrids that has notes of strawberry and honeydew melon.

Magnum (13–16% alpha acid): Another non-noble hop, Magnum has become the bittering hop of choice for many brewers on both sides of the Atlantic. It has a clean, smooth bitterness, and its high alpha percentage means less hop material in the kettle (and less money per batch) compared to noble hops.

Mandarina Bavaria (8–10% alpha acid): One of three new German hybrids bringing New World character into traditional hops, Mandarina has tangerine and orange notes emphasized by dry hopping.

Perle (7–9% alpha acid): This isn't a noble hop, but it's used widely for bittering, even in famous ales such as Sierra Nevada Pale Ale.

Saaz (3–4.5% alpha acid): Saaz is the classic Czech (Bohemian) lager hop. It has a delicate, fruity aroma and taste. It also has a smooth bitterness. It's essential for Bohemian pilsners, such as the classic Pilsner Urquell. American-grown Saaz isn't a good substitute, although American Sterling is—especially if you're making an aggressive pilsner.

Spalt (4–5.5% alpha acid): This is a good all-purpose hop. It isn't used as widely as other German hops, however, and can be harder to find. It's the classic hop for brewing Düsseldorf Alt beers.

Tettnang (3.5–5.5% alpha acid): This is a popular all-purpose hop for any German beer. Its spicy, floral notes make for a unique pilsner, but it's also fantastic in wheat beers or a Munich Helles.

Other Hops
Australian and New Zealand Hops

Hops from down under are finding their way into more and more North American breweries. Many have unique aromas and tastes that people really enjoy. These haven't been used widely due to a lack of consistent supply, but some small and midsize breweries do use them. Galaxy and Nelson Sauvin™ in particular have stood out in recent years. The former brings intense citrus and passion-fruit taste, while the latter also is fruity but with some white-wine notes. Pride of Ringwood is an old-school Australian hop with none of the funk or flavor of new varieties. Its earthy aroma and taste are polarizing and likely best avoided.

Galaxy (12–16% alpha acid): One of the most coveted new hops, this Australian beast is oily and sticky with an aroma of pine, citrus, cannabis, and ripe tropical fruit. Fantastic as a dry hop, especially in funky Brett beers.

Kowatu™ (6–7% alpha acid): One of my favorites of the New Zealand hops, this has loads of tropical fruit combined with a bright pine note. Great for dry hopping.

Motueka™ (6.5–7.5% alpha acid): One of the mellower New Zealand hops with light citrus and tropical fruit aromas. Best used in lighter beers to let the hop shine.

Nelson Sauvin (12–13% alpha acid): Named after the Sauvignon Blanc grape, this variety is one of the most sought New Zealand hops in the craft beer scene. Big aromas of white grape, lychee, and mango leap from the glass. Great for dry hopping.

Rakau™ (6–10% alpha acid): This fairly restrained New Zealand variety has aromas of stone fruit (peaches and apricots) and is best used in light beer styles that let the hop shine.

Riwaka™ (4–6% alpha acid): Descended from Saaz hops, this New Zealand variety has a big juicy aroma of kumquats and citrus.

Vic Secret™ (14–17% alpha acid): A juicy, tropical aroma makes this Australian hop a great choice for dry hopping modern IPAs.

Wai-iti™ (2.5–3.5% alpha acid): This delicate, low-alpha variety features notes of lime zest and peach.

Waimea™ (16–19% alpha acid): Maybe my favorite New Zealand variety, this one pairs a monster level of alpha acid and oil with a huge tropical-fruit and pine aroma. Great for dry hopping.

Wakatu™ (6.5–8.5% alpha acid): This new Zealand varietal has pleasant floral and lime zest aromas.

European Transplants
Examples include Crystal, Liberty, Santiam, Sterling, Ultra, and Willamette
This handful of hops was bred to emulate European hops. Do they smell and taste exactly like European hops? Not really, but they're all interesting with pretty low alphas (3–5 percent). If you're trying to reproduce a traditional European beer style, don't use these. If you're looking for a unique take on a style or an accent hop to blend with another American hop, give them a shot.

Less-Hyped American Hops
Just because a hop isn't highly sought doesn't mean it's not worth trying. Simcoe was around for years, and hardly anyone used it! By the time it became a brewers' favorite, it almost had disappeared. Many breweries use lesser-known hops as secret weapons. Examples include Crystal, Glacier, Mount Hood, and Newport. They may serve as midboil additions, where they act as a nice flavor hop without using precious hops such as Simcoe and Amarillo. Or breweries may craft a dry-hop blend that uses some of the bigger names but balances them with other hops. Whether you want to see what they bring to a beer with a SMaSH recipe (page 42) or just experiment on the fly, you won't have your own secret-weapon hops until you experiment with lesser-known varieties. Try checking the websites of your favorite breweries if you want to decode some of the hops you smell and taste in their beers. You might find one you like.

Hop Blends
Some hops are blends created by hop distributors. Examples include C-Type Blend, Falconer's Flight, and Falconer's Flight Seven C's. They're formulated specifically to be, for example, a citrusy blend or a tropical-fruit blend and often include unnamed, experimental varieties. Hop blends will be similar from year to year but are never exactly the same. Also, if a blend becomes unpopular or if demand for it declines, it could cease to exist a lot faster than an established hop.

1868 EAST INDIA PALE ALE

That isn't a typo below: You add about 13 ounces of Fuggle hops at the start of the boil! Then you dry hop with an additional 2½ ounces of Fuggle. Buying hops by the pound is much cheaper and, for this recipe, a no-brainer. This recipe calculates out at 172 IBUs, but it produces a crisp, resiny, golden-colored ale with immense complexity. My tastebuds tell me it's around 85 to 95 IBUs.

YOU NEED
basic brewing equipment (page 3)

12½ gallons filtered brewing water (page 12)

17 pounds Maris Otter pale-ale malt (100%)

59 alpha acid units Fuggle hops at 90 minutes (172 IBU, calculated)

2½ ounces Fuggle hops (dry hop)

fine-mesh strainer

1 Whirlfloc tablet

2 packages London Ale III WY1318 yeast (or a 2-liter starter made from 1 pack, page 110)

3¾ ounces dextrose/corn sugar (optional, use only for bottling)

TARGETS
Yield: 5 gallons
OG: 1.067–1.069
FG: 1.012
IBU: 172 (calculated)

Note: This is a 6½-gallon batch, which allows for 5 gallons of finished beer after hop absorption in the kettle and fermentor.

1. Mix the malt with 6 gallons of water at 165°F or the appropriate temperature to mash at 150°F. Mash for 60 minutes.

2. Recirculate the wort until it's fairly clear. Run off the wort into the kettle.

3. Sparge with 6½ more gallons of water at 165°F. Run off the wort into the kettle.

4. Bring the wort to a boil and add the hops. Don't be afraid—do it! After 60 minutes, use a fine-mesh strainer to remove as much hop material as possible. Keep running the strainer through the wort and disposing of the collected hop residue. At the end, you'll have a pile of spent hops the size of a bowling ball.

5. Once you're finished, add the Whirlfloc tablet and continue to boil for 20 minutes. Put your wort chiller into the wort at least 15 minutes before the end of the boil.

6. When the boil finishes, cover the pot with a lid or a new trash bag and chill to 68°F. Siphon the wort into your sanitized fermentor and pitch two packs of liquid yeast or a 2-liter starter.

7. Ferment at 68°F for one week, then let it warm to 70°F for the second week. Add the dry hops when fermentation slows down (usually after five to seven days).

8. Keg or bottle the beer. (If you're bottling, I recommend 3¾ ounces of dextrose/corn sugar for this beer.)

HOPPING TECHNIQUES

TRADITIONAL HOPPING

Homebrewers originally emulated the basic hop schedule of many commercial breweries. It goes like this: Add the amount of hops you need to get the bitterness you want at the start of the boil, add a small amount of hops 20 minutes before the end of the boil for hop flavor, and throw in another small addition at the end of the boil for hop aroma. This strategy can make fine beer. But it doesn't quite emulate commercial breweries thanks to the whirlpool rest that commercial breweries perform. (See "Hop Bursting," page 88).

MASH HOPPING

I remember when the idea of adding hops with the grain was introduced to the homebrew community. I didn't understand how hops could contribute bitterness if they never got above 160°F. So I did an experiment where I added 6 ounces of Cascade hop pellets into a simple mash of pale malt. No other hops were added throughout the boil. While I wish I could say I was blown away, the resulting beer had no notable bitterness or aroma (as expected). The mash temperature was too low to isomerize the alpha acids in the hops. The exception would be in a decoction mash, where you bring the grain and the hops to a boil. Old German brewing texts refer to this technique as "hop roasting."

FIRST-WORT HOPPING

Homebrewers jumped on this technique after reading about it in a German technical publication. First-wort hopping (FWH) involves adding your hops to the wort as you start to collect it in the kettle. The hops soak into the wort until you collect the total amount and bring it to a boil. Then you continue the boil as usual. The original publication stated that professional tasters preferred a beer brewed using FWH (as opposed to adding the hops after the wort had already come to a boil) and claimed that it gave the beer a smoother bitterness. It didn't take homebrewers long to assert that it gave great hop aroma as well! Most modern brewing software calculates the bitterness addition of first-wort hops as similar to that of a 20-minute addition. In practice, I didn't find that to be true. See the experiment at right.

FIRST-WORT HOPPING EXPERIMENT

For this experiment, I brewed two batches of a typical pale ale. To the first batch, I added 40 IBUs' worth of hops at the start of the boil. To the second batch, I added the same amount of hops at first wort. No other hops or mineral additions were made. According to those who claim the first-wort hopping would give the bitterness only of a 20-minute addition, my IBUs should total 24.

Lab analysis showed however that the first wort-hopped beer had 39 IBUs (reassuringly close to the target of 40), while the traditionally hopped beer had 38 IBUs—close enough to assume that the same amount of IBUs exists in both first-wort and traditionally hopped beers.

A blind tasting panel of BJCP judges had difficulty finding much difference between the two samples, but they slightly preferred the traditionally hopped beer over the first-wort hopped beer. They described the traditional beer as being less harsh and having a bit more hop character.

The takeaway from this experiment is that you should calculate first wort-hops not as a 20-minute addition but as a start-of-the-boil addition. Any dreams of a great hop aroma from FWH are probably still just dreams.

HOP BURSTING (WHIRLPOOL HOPPING)

When commercial breweries turn the heat off on the kettle, they do what's called a "whirlpool rest." Usually the wort stays in the kettle, but some breweries have a separate whirlpool tank. Either way, a pump gets the wort spinning, which usually takes 10 to 15 minutes. Then the pump is shut off, and the brewer waits another 15 minutes for the wort to stop spinning. This process deposits most of the hop residue and break material in the center of the tank, so the wort that runs into the fermentor is clearer than it would have been without the whirlpool. The wort then goes through a heat exchanger (wort chiller), which can take another 60 minutes.

These extra steps mean that the hops that a commercial brewer adds at knockout (when the heat is turned off) are sitting in near-boiling wort for at least 30 minutes and up to 90 minutes. Even though the wort isn't boiling vigorously, it's still hot enough to extract bitterness and aromatic oils from the hops. This can wreak havoc if you're trying to calculate hop bitterness in

a homebrew recipe. Normally end-of-the-boil additions don't add any bitterness, but a hop addition sitting near boiling for 45 minutes can provide significant bitterness.

Knowing this, homebrewers started experimenting by adding more and more hops at the end of the boil and letting it sit hot for 30 to 60 minutes. The prevailing wisdom (and my own experience) shows that you'll get around a third to a half of the bitterness that you'd get with a start-of-the-boil addition. So if you normally add 1 ounce of a 15 percent alpha acid hop for bittering an IPA, you want to add 2 to 3 ounces of the same 15 percent alpha acid hop at the end of the boil and then wait 45 minutes before starting to chill. This should get you the same level of bitterness but with retaining more of those volatile hop oils that normally boil off.

The technique of adding most or all of the hops at the end of the boil is called "hop bursting" or a "hop stand." The goal is big hop aroma and taste with a rounded bitterness. (See the Hop-Bursting Pale Ale recipe, page 92, for an example.) Of course, you also can combine hop bursting with a small bittering addition.

DRY HOPPING

Tossing an ounce or two of hops into a beer near the end of fermentation and letting them sit in there for a few days will give you a bright, fresh hop aroma. This process is known as dry hopping.

In the past, British brewers were the main proponents of dry hopping. They added a few handfuls of hops into a cask before bunging it up. But over the last decade, home-brewers and commercial brewers in America have been pushing the envelope of hop aroma, experimenting with any technique that will provide something more intense. This ongoing quest has produced a wealth of variations on dry hopping for maximizing aroma and flavor.

Basic Dry Hopping

After the primary fermentation has finished (five to seven days), throw anywhere from $1/2$ to 6 ounces of hops (preferably pellets) into the beer. Allow it to sit at room temperature for three to seven days to absorb the aromatic hop oils. Some brewers think a shorter time results in better aroma; others think a longer time is key to more hop character. I like three to five days of dry hopping, and you can see why on the next page. If you can chill the fermentor before transferring the beer to a keg or bottling bucket, the pellet residue will settle to the bottom of the fermentor (especially if you use a fining such as gelatin, page 123).

Note: When you're brewing an imperial IPA or other big beers that need three to four weeks of aging to mellow the alcohol, it's best to add dry hops after three weeks of aging in order to retain the most aroma.

Dry Hopping Warm vs. Cold

In the past, homebrewers preferred to cool their beer after fermentation and then dry hop, mainly to emulate what commercial breweries were doing. At a commercial brewery, once beer has finished fermentation, it's usually given an additional day or two to allow the yeast to "clean up" any fermentation off-flavors. It's then crashed to 32°F to flocculate the yeast cells (force them to sink to the bottom) and to form chill haze, which then will be filtered out.

But there's a drawback to dry hopping after the beer has been crashed. Hop oils don't absorb nearly as well into solution as they do at fermentation temperatures. The Catch-22 is that, until you crash the beer, the yeast takes much longer to settle out. So brewers had a conundrum: Dry hop warm with the yeast in solution and get better oil extraction, or dry hop cold so the hop oils aren't affected by the yeast. The consensus these days is that dry hopping warm yields the best results, even though yeast will drag some of the hop oils out of solution.

HOP BIOTRANSFORMATION

On the cutting edge of hop aroma research is a new technique called "hop biotransformation." In layman's terms, yeast is interacting with the hop compounds and turning them into something else. Instead of the traditional method of dry hopping, in which fermentation ends and most of the yeast has dropped out, this new technique advocates dry hopping in the middle of fermentation to have the hops and yeast interact.

This method is very popular with brewers of the new NE IPA style.

I haven't experimented with this new technique enough to say much about it, but it looks promising as a way to create new aromas from hops. If you belong to a homebrew club, consider having two brewers brew the same beer with the same hops but using the different dry-hop techniques. Then have a blind tasting.

Some commercial breweries now go through the trouble of crashing the beer for a day or two to flocculate the yeast, then letting the beer warm back up to fermentation temperatures to dry hop. This works best when the fermentation vessel allows for the removal of yeast from the bottom (such as a conical fermentor). In practice, the slightly higher utilization rate probably won't make a huge difference in a 5-gallon batch. Homebrewers, who as a rule are less concerned about the bottom line, can add a pinch more hops to achieve a similar effect.

Dry Hop Contact Time

If you ask 10 brewers how long you should dry hop, you'll probably get 10 different answers. Even among pros known for their hoppy beers, the contact time frequently ranges from 3 days to 14. To get some data on different contact times, I split a batch into four kegs and dry hopped each for a different length of time. The goal was to find out if there was a sweet spot where waiting any longer didn't add more aroma.

Recipe and Process

I brewed the Malverde IPA found on page 96. After fermentation, I crashed the beer to 40°F overnight to settle the yeast, then split the clear beer into four CO_2-purged kegs and allowed them to warm back up to 65°F. Over the next two weeks, I added hops to each keg. These were the pellet hops for each 3-gallon batch: 2/3 ounce Simcoe, 2/3 ounce Columbus, 2/3 ounce Centennial, and 2/3 ounce Citra.

Keg 1 was dry hopped for 14 days total.
Keg 2 was dry hopped for 7 days total.
Keg 3 was dry hopped for 5 days total.
Keg 4 was dry hopped for 3 days total.

At the end of 14 days, all beers were crashed, transferred into fresh CO_2-purged kegs, and carbonated. The beers then were tasted blind by BJCP judges and fellow brewers in random orders. The results were clear: The vast majority of tasters thought that the beer dry hopped for 3 days had the hoppiest aroma. Several also liked the beer dry hopped for 5 and 7 days, which they thought had a bit more hop taste than the 3-day dry hop. (Almost all tasters thought that the 5- and 7-day beers were identical.) Last and perhaps most importantly, nearly every taster thought that the 14-day batch had the least appealing hop aroma and taste. The take-home lesson here is that at 3 days you have absorbed most of the aromatic oils from the hops that produce aroma (at least when using pellets).

Multiple Rounds of Dry Hopping

Not satisfied with a single addition of dry hops, some hop-crazed brewers use a staggered approach. They start by adding some dry hops during the last day of active fermentation. After waiting around 3 days, they add another dose of hops. Then, after an additional 3 days, they add even more dry hops. The logic behind this method is that different components of hops are extracted depending on how much yeast is in solution. The brewers are hedging their bets by staggering the hops throughout the clarification process.

Dry Hopping in the Keg

Adding hops into a clear, finished beer as it's being kegged is the traditional approach for British cask beer. The benefit to this late hop addition is that the hop aroma is trapped in the keg—and there's little yeast to inhibit hop oils from staying in solution. One downside is trying to contain the hops. Putting pellets in a nylon bag results in minimal surface contact, and leaving any form of hop loose in a keg will inevitably clog up the works. Using whole hops loosely packed into muslin bags works well, but keep in mind that the hop character will be constantly changing over the weeks that you're consuming the beer. Some homebrewers caution that the beer may take on a grassy flavor if the hops are left in for too long, but I haven't found that to be the case.

HOP-BURSTING PALE ALE

This beer has a clean, simple malt bill that lets the hop bursting take center stage (page 84). To emulate the brewery whirlpool effect, after you add all the hops at the end of the boil, let them sit for 45 minutes before chilling. That way, the hops will isomerize slowly while the aromatic oils are extracting. Pinpointing the IBU level you'll get from adding the hops this way can prove difficult. Brewing calculators say 0, but my experience says that it should be around 40 to 60 IBUs.

YOU NEED

basic brewing equipment (page 3)

9 gallons filtered brewing water (page 12)

10 pounds North American 2-row malt (90.9%)

1 pound medium crystal malt (40°L) (9.1%)

1 Whirlfloc tablet

24 alpha acid units Chinook hops at 0 minutes

24 alpha acid units Citra hops at 0 minutes

15 alpha acid units Columbus hops at 0 minutes

2 vials or packages California Ale WLP001/ American Ale WY1056 yeast (or a 2-liter starter made from 1 pack, page 110)

3¾ ounces dextrose/corn sugar (optional, use only for bottling)

TARGETS
Yield: 5 gallons
OG: 1.055–1.057
FG: 1.012
IBU: likely 40–60

1. Mix the malt with 4 gallons of water at 165°F or the appropriate temperature to mash at 150°F. Mash for 60 minutes.

2. Recirculate the wort until it's fairly clear. Run off the wort into the kettle.

3. Sparge with 5 more gallons of water at 165°F. Run off the wort into the kettle.

4. Bring the wort to a boil. Boil it for 75 minutes. Add the Whirlfloc tablet at 30 minutes. Put your wort chiller into the wort at least 15 minutes before the end of the boil. When the boil finishes, turn off the heat and add all the hops, but don't start to chill the wort yet. Cover the pot with a lid or a clean trash bag and wait 45 minutes. Then chill to 65°F.

5. Siphon the wort into your sanitized fermentor and pitch two packs of liquid yeast or a 2-liter starter.

6. Ferment at 65°F for one week, then let it warm to 68–70°F for the second week.

7. Keg or bottle the beer. (If you're bottling, I recommend 3¾ ounces of dextrose/corn sugar for this beer.)

NE JUICY IPA

It's rare for a style to appear almost from nowhere and take the beer world by storm, but that's exactly what has happened with NE (New England or Northeast) IPAs. What makes them different from a typical West Coast IPA? One glance will tell you that it even looks different. These new IPAs embrace and even strive for a strong haze that sometimes borders on murky. The huge aroma focuses on citrus and tropical-juice character. The bitterness tastes more subdued than that of its West Coast sibling, and it has a much fuller body. Some breweries even add lactose to sweeten the beer to add to the "juice" aspect.

YOU NEED

basic brewing equipment (page 3)

12½ gallons filtered brewing water (page 12)

14 pounds English pale ale malt

4 pounds flaked oats

14 alpha acid units Columbus hops at 60 minutes (38 IBUs)

8 ounces Citra, Galaxy, Waimea, or Nelson Sauvin hops (end of boil)

2 packages London Ale WY1028 or WLP013 yeast (or a 2-liter starter made from 1 pack, page 110)

8 ounces Citra, Galaxy, Waimea, or Nelson Sauvin hops (dry hop)

4.4 ounces dextrose or corn sugar (optional, use only for bottling)

TARGETS
Yield: 5 gallons
OG: 1.066–1.070
FG: 1.014
IBU: 40–50

Note: This is a 6½ gallon batch, which allows for 5 gallons of finished beer after hop absorption in the kettle and fermenter. This beer is all about freshness and should be consumed within a few weeks of packaging. Kegging is recommended if possible. Also, if adjusting your mineral content, use the Malt-Forward Water Profile (page 59).

1. Mix the malt with 6 gallons of water at 165°F or the appropriate temperature to mash at 150°F. Mash for 60 minutes.

2. Recirculate the wort until it's fairly clear. Run off the wort into the kettle.

3. Sparge with 6½ more gallons of water at 165°F. Run off the wort into the kettle.

4. Bring the wort to a boil and add the hops. Put your wort chiller into the wort at least 15 minutes before the end of the boil.

5. When the boil finishes, cover the pot with a lid or a new trash bag and add the 8 ounces of whirlpool/knockout hops. Let sit for 5 minutes.

6. Chill to 68°F. Siphon the wort into your sanitized fermenter. Don't worry about leaving hop residue behind; there's going to be a lot.

7. Pitch two packs of liquid yeast or a 2-liter starter.

8. Ferment at 68°F for one week, then let it warm to 70°F for the second week. Add the dry hops when fermentation slows down (usually after five to seven days.) For maximum haze, don't fine with gelatin. Cold crash it overnight to drop the hops out.

9. Keg or bottle the beer. (If you're bottling, I recommend 4.4 ounces of dextrose/corn sugar for this beer.)

IPA THREE WAYS

I love, love, love IPAs. New England, West Coast, Burton-on-Trent, black, white—I love them all. A lot of variations exist among IPAs in terms of color and malt profile, but the basic recipe for all of them is about the same: an OG of 1.062–1.067 and an IBU of 60–70, not too heavy on the crystal malt. Brewing a good IPA isn't rocket science, but making a perfect one is a never-ending quest. After more than 100 attempts, I've found that the recipe below is pretty close to perfect (for me, and for now). It includes several variations if you want to get into black or white IPAs.

MALVERDE IPA

YOU NEED

basic brewing equipment (page 3)

12 gallons filtered brewing water (page 12)

16 pounds North American 2-row malt (94%)

1 pound German melanoidin malt (6%)

22 alpha acid units Columbus hops at 75 minutes (61.9 IBU)

1 Whirlfloc tablet

5.5 alpha acid units Centennial hops at 15 minutes (5.8 IBU)

6 alpha acid units Simcoe hops at 5 minutes (3 IBU)

1/2 ounce each Simcoe, Citra, and Centennial hops at end of boil

1 ounce each Simcoe, Columbus, Citra, and Chinook hops (dry hop)

2 vials or packages California Ale WLP001/ American Ale WY1056 yeast (or a 2-liter starter made from 1 pack; page 110)

3¾ ounces dextrose/corn sugar (optional, use only for bottling)

TARGETS
Yield: 5 gallons
OG: 1.064–1.068
FG: 1.011
IBU: 70

Note: The recipe is for 6½ gallons. You'll have a lot of hop residue in the kettle, and you want to get at least 5½ gallons of finished wort going into the fermentor. Once you dry hop with 4 ounces of hops, you're going to lose another ½ gallon of beer.

1. Mix the malt with 5½ gallons of water at 165°F, or the appropriate temperature to mash at 150°F. Mash for 60 minutes.

2. Recirculate the wort until it's fairly clear. Run off the wort into the kettle.

3. Sparge with 6½ more gallons of water at 165°F. Run off the wort into the kettle.

4. Bring the wort to a boil, add the first addition of hops, and boil for 75 minutes. Add the Whirlfloc tablet with 30 minutes left in the boil. Put your wort chiller into the wort at least 15 minutes before the end of the boil. Add the hop additions as listed at left.

5. When the boil finishes, cover the pot with a lid or a new trash bag and chill to 65°F. Siphon the wort into your sanitized fermentor and pitch two packs of liquid yeast or a 2-liter starter.

6. Ferment at 65°F for one week. Let it warm up to 68–70°F during the second week, adding the dry hops after primary fermentation slows down (usually after five to seven days).

7. Keg or bottle the beer. (If you're bottling, I recommend 3¾ ounces of dextrose/corn sugar for this beer.)

VARIATIONS

Black IPA: Substitute 1 pound of German Carafa III, Briess Black Prinz, or Midnight Wheat malt for 1 pound of the 2-row malt. Everything else stays the same.

White IPA: Change the grain bill so that half of the 2-row is wheat malt. Eliminate the melanoidin malt and add ½ pound of rice hulls. Everything else stays the same.

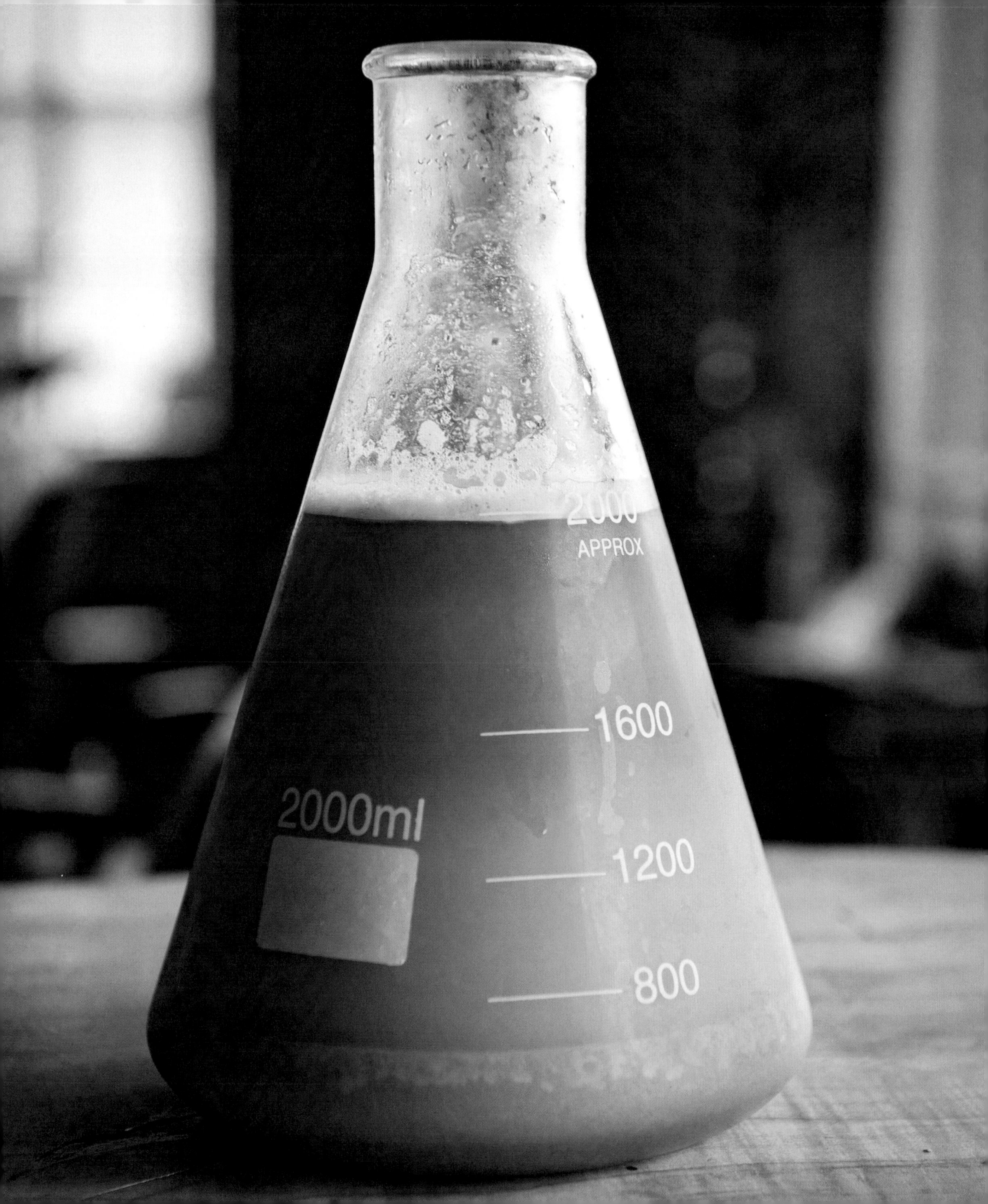
APPROX
1600
2000ml
1200
800

YEAST and FERMENTATION

YEAST QUALITIES AND SELECTION

WYEAST
Direct Pitch ACTIVATOR™ for Brewing

3763 ROESELARE
MFG 06 NOV 12

100 Billion Yeast Cells

The Direct Pitch **ACTIVATOR**™ contains 100% Pure Liquid Yeast plus an internal nutrient packet. This Wyeast **Smack-pack**™ system allows yeast metabolism to begin prior to inoculation as well as providing proof of yeast health. Our **ACTIVATOR**™ is designed to inoculate 5 gallons of wort with the same pitch rate recommended by professional brewers.

Contains: 125 mls (4.25 Fl. Oz. or 128 grams) yeast and nutrients

wyeastlab.com

Selecting a yeast often starts with matching its attributes to your recipe—fruity or clean? sweet or dry?—you also must consider factors such as temperature tolerance, alcohol tolerance, and flocculation to determine whether a yeast strain suits your needs. Homebrewers have an advantage over commercial brewers when it comes to yeast selection: A homebrewer can choose a specific strain for each batch of beer, whereas commercial brewers usually have to limit themselves to just a few strains. For breweries, this has to do with convenience as well as expense. With one house strain, there's always fresh yeast to harvest around the brewery—and you get to know that strain extremely well. Plus, a pitchable amount of a specialty strain could cost a brewery close to $1,000! (What you pay for yeast doesn't seem so bad in comparison, does it?)

Most breweries use pure liquid cultures bought from the same companies that sell to homebrewers: Wyeast and White Labs. For homebrewers, Wyeast packages their yeast in a foil pouch with a nutrient pack inside. When popped, it gives the yeast some food and nutrients that cause the package to swell. While it's no substitute for making a starter, it does let you know that the yeast is alive in there. Even a slight swelling of the package is OK, according to Wyeast. White Labs, on the other hand, packages their yeast in thick plastic vials, with no built-in nutrient pack. Both companies are stellar in quality. If given the choice, pick the fresher of the two.

BOUTIQUE YEAST COMPANIES

One of the most exciting developments of modern homebrewing is the proliferation of tiny yeast labs offering local and exotic strains. See the end-of-book resources (page 190) for a list of boutique yeast labs and some of my favorite strains that they produce. Because of the limited availability of these products, this book focuses on the yeast products that your local homebrew shop most likely will carry.

Note: In this book, when I talk about a specific yeast strain, I give names and product numbers for both companies, like this: California Ale WLP001/American Ale WY1056. White Labs yeast will appear first, accompanied by three numbers. Wyeast's yeast will follow, along with four numbers. (Some strains are unique to one particular manufacturer and will not have an equivalent.)

ESTERS AND PHENOLS

Esters are fruity aromas and tastes created by yeast at the beginning of fermentation. Each strain produces different amounts of various esters that give it its personality. Some strains, such as California Ale, produce very low levels of esters and are considered clean yeasts. However, a yeast that produces no esters at all makes a very bland beer—so some esters are good!

The amount of esters produced by yeast increases with temperature, beginning at the time the yeast is pitched. This is why it's important to pitch cool and keep the fermentation temperature under control for the first few days. If the beer starts fermenting too warm, it doesn't help to cool it down later—the damage has been done already. Insufficient levels of oxygen at the beginning of the fermentation also can increase esters to an undesirable level.

These are the most common esters and their signature aromas:

➡ Isoamyl acetate: banana
➡ Ethyl acetate: solvent or nail polish remover
➡ Ethyl caproate: apple

Unlike esters, phenols usually are considered an off-flavor in beer. Some common descriptors are plastic, smoke, or model-airplane glue. Usually they're caused by a wild yeast infection, because most traditional brewing yeasts lack the ability to create phenols. (An exception: German wheat beer strains and Belgian strains, which can produce the phenol 4-vinyl guaiacol, give hefeweizens their characteristic clove aroma and taste.) When a yeast has the ability to create phenols, it uses ferulic acid in the wort to create them. Brewers of hefeweizen often perform a mash rest at around 114°F, known as a ferulic acid rest. This creates more ferulic acid, which, in theory, can increase the levels of the phenol and, therefore, add more clove character.

DIACETYL

One of the most common faults in homebrewed and commercial beers is diacetyl, caused by the conversion of an ester released as a by-product of fermentation. Diacetyl's aroma and taste are so similar to butter that some food manufacturers use it for butter flavoring. Diacetyl can get into your beer in two ways: through a wild yeast infection or through a normal yeast strain that wasn't allowed to clean up after itself.

Any yeast strain can produce diacetyl, but most will reduce the levels to below the taste threshold if given enough time and a warm enough temperature. If a strain is notorious for producing large amounts of diacetyl or needs more time than most yeasts to clean up diacetyl, the strain description from the manufacturer will recommend a diacetyl rest. A diacetyl rest warms the beer higher than the fermentation temperature as fermentation is finishing, then holds it at the warmer temperature for a few days. This helps the yeast reabsorb and break down the diacetyl. For ale yeasts, warm the beer about 5–10° higher than your fermentation temperature. For lagers, raise them all the way up to 65–70°.

If you're worried about diacetyl in a beer, you can do a forced diacetyl test. Take out a small amount of beer, put it in a small glass jar covered with foil, place it in a hot water bath (about 160°F), and let it sit for an hour. Then chill the sample before you smell and taste it. If there's no buttery flavor, then you can assume that the beer is fine. If a bottled beer has signs of diacetyl, you can determine easily whether it's an infection—bad haze and a ring around the neck of the bottle usually accompany the buttery aroma. If the beer is clear and there's no ring, chances are you need to perform a diacetyl rest the next time you use that particular yeast.

TEMPERATURE TOLERANCE

A fermentation chamber, such as a chest freezer with a temperature regulator, is one of the best investments you can make when you get serious about brewing. Once you have the ability to set and control temperatures, you have the ability to use any yeast. It's still important, however, to read up on your yeast and follow a fermentation schedule suited to it. (See page 115 for more on temperature schedules for fermentation.) Once you dial in a yeast strain, you can re-create fermentation environments—and consistently make the same beer without depending on the whims of the weather or your home's thermostat.

If you don't have a fridge or freezer with a temperature regulator, then you're at the mercy of ambient temperatures in your house or basement. But there's nothing wrong with brewing by season. In warmer months, spend time exploring saisons and other beers with Belgian strains of yeast,

which generally produce better aromas and tastes with temperatures above 70°F. (Some work better at very high temperatures.) If it's not too hot where you live, you can push some strains of clean American ale yeasts into warmer-than-normal temperatures, which will give you fruitier ales that are still tasty. Here's a sampling to get your ideas flowing.

California Ale WLP001/American Ale WY1056: So many breweries use Cal Ale for their house yeast because it's a workhorse that can ferment down to 58°F if you want a clean, slightly sulfury pseudo lager. It can also ferment to 80°F and still produce a fine, if slightly estery, ale. When I use it, I start the fermentation at 65–67°F and let it warm up to 71°F for the last few days of fermentation. Cal Ale produces very little diacetyl but can take a while to clear up. It responds well to clarifiers such as gelatin if you're looking for clearer beer (page 123).

Belgian Saison WLP565/WY3724: This strain can be problematic if fermented too cool—meaning below 75°F! That makes it a great strain for the dog days of summer. I've brewed fantastic saisons with this yeast by fermenting at 95°F. The biggest mistake you can make with this yeast is not to check the final gravity. It can look completely finished, but the gravity might still be 1.030. Give the fermentor a swirl and be patient—it'll keep going and going. The final gravity should fall below 1.010 and sometimes as low as 1.003.

Abbey WLP530/Trappist WY3787: Three traditional Trappist breweries use this strain: Westmalle, Westvleteren, and Achel. Each brewery produces distinct beers by fermenting at different temperatures. Rumor has it that Westmalle keeps it at 68°F and Westvleteren lets it rise to 84°F. It's a fun yeast to play around with and get to know. At low temperatures, it's relatively clean, so a complex malt bill will shine. At high temperatures, it has increased dark-fruit flavors that work best with a simple malt bill.

Super Yeast WLP090: This is a great house strain if you want a clean flavor with better flocculation than Cal Ale. It's a quick, strong fermenter that can knock out a beer in four days! That isn't a huge selling point for homebrewers, but for a commercial brewery, it can mean a huge increase in production. It ferments best around 65–68°F, but since it's so quick you could take advantage of a cool week in your part of the country. Keep in mind that the vigorous fermentation can mean higher-than-usual heat buildup in the fermentor.

ALCOHOL TOLERANCE

If you're making a beer with an alcohol content above 10% ABV, then you need to choose an appropriate yeast. Almost any yeast strain can handle up to 10 percent alcohol, but for anything above that you need to choose a yeast that can keep working in a high-alcohol environment. While strain is important, it's even more important to make sure you're pitching the proper amount of yeast, adding the appropriate amount of oxygen, and driving the fermentation through temperature control.

The Super Yeast mentioned above would be a great choice for most beers, as would good old Cal Ale. I've gotten Cal Ale up to 15% ABV, but it has problems going much higher. White Labs has a strain specifically for high-alcohol beers that supposedly can go up to 25% ABV, but it's very needy and difficult to work with, and some brewers haven't liked the overall flavor. Homebrewers often hear that they should add Champagne yeast to a stuck fermentation, but Champagne yeast and wine yeast in general can't ferment the long-chain sugars in beer and so won't be of much use to brewers.

Other alcohol-tolerant strains include Dry English Ale WLP007, just about any Trappist or Abbey yeast, Zurich Lager WLP885, and Denny's Favorite 50 WY1450. Never repitch yeast from a high-alcohol brew (more than 7% ABV), because it likely will be too stressed to perform well again.

FLOCCULATION

Highly flocculent strains drop from solution more than lower flocculent strains, but flocculation actually measures the degree to which the yeast cells stick or clump together. The more they stick together, the heavier they get and the faster they drop out and settle to the bottom of the tank.

Highly flocculating yeasts often leave more residual sugars behind, which can result in a fuller-bodied beer. But they also can leave more diacetyl behind compared with a low-flocculating yeast, which hangs out in solution and reabsorbs diacetyl. Low levels of calcium can slow the flocculation of yeast, so if you check your water (page 56), make sure you always have at least 50 ppm.

ATTENUATION

Attenuation measures the percentage of fermentable sugars that yeast converts during the fermentation process. Most yeasts have attenuation in the 70 to 75 percent range. To figure out your approximate finishing gravity, take the last two numbers of your OG (1.060 = 60) and divide that number by 4 (60 ÷ 4 = 15). Put that number back in its place (1.015), and you have your final gravity if your yeast gets 75 percent efficiency. Just as with alpha acids and other brewing numbers, online calculators are available and highly recommended for more accurate estimations. See page 191 for recommendations.

The attenuation rates for any yeast are available from the yeast manufacturer's website, but take them as more of a general guideline than hard fact. Mash temperature, grain bill, and other factors can affect attenuation. Yeast can consume only fermentable sugars, so if the wort has a lot of unfermentable sugars, the attenuation will be lower. Generally, very flocculant yeast also has lower attenuation because it settles out of solution more quickly. This can sometimes be fixed by rousing the yeast back into solution by swirling the fermentor, but if you're worried about attenuation for a particular beer, it's probably better to use a less-flocculant strain.

COMMON TYPES OF YEAST

Saccharomyces is the genus of yeast used for the majority of brewing, and was probably the earliest domesticated organism. Egyptian ruins from 4,000 years ago depict the yeasting of bread. The original yeasts were undoubtedly wild, present in the air or on the grain and fruit used to make primitive, ancient beers. By continuing to select the best-tasting yeasts and discarding the ones that tasted unpleasant, humans unknowingly selected the first brewing-yeast cultures!

The top-fermenting/bottom-fermenting method of separating ale and lager strains—meaning that ale yeast rises to the top of the beer, and lager yeast sinks to the bottom—is overly simplistic. Plenty of lager yeasts form a large krauesen (the foamy head on the beer during fermentation), and plenty of ale yeasts don't. Even the rule that ale strains ferment warm (65–70°F) and lagers ferment cold (47–52°F) is flawed, since some ale yeasts can ferment down into lager temperatures and vice versa. However, that's typically the way yeast manufacturers divide yeasts, so we'll stick to it for now.

Technically, ale yeast strains are members of the *Saccharomyces cerevisiae* species, and they do tend to be warm fermenters, shutting down and refusing to ferment at temperatures below 55°F. This is a motley crew of yeasts, ranging from super-clean yeasts to spicy, estery Belgian and fruity British strains.

Lager yeast taxonomy seems to change every decade. It used to be known as *Saccharomyces carlsbergensis*, then became *Saccharomyces uvarum*, and now is *Saccharomyces pastorianus*. Lager yeast can continue fermenting below 40°F, although most lager fermentations happen in the 48–52°F range. Lager yeasts can ferment complex sugars that ale yeasts can't, but these sugars exist in such tiny amounts in wort that the difference in attenuation isn't notable.

Countless strains are now available, and it isn't possible to cover all of them here. As a brewer, you're going to need to do some experimentation with different strains. Taste them yourself and get to know their individual personalities. By familiarizing yourself with some of the most popular strains used by homebrewers and commercial brewers alike, you'll get an idea of what yeasts you might want to try for a particular beer. From there, brew up a batch, split it between two yeasts, and compare the different yeast characters.

USING TWO STRAINS OF YEAST

Back in the day, most breweries used a multistrain culture of yeast. Today, brewers use multiple yeasts for three main reasons:

1. Blends for flavor: Yeast labs today sell special blends of different cultures to achieve certain complex flavor profiles. Using a blend or multiple yeasts means that, if you repitch, one strain starts to dominate. Over time, the flavor will change.

2. Tag-teams for attenuation: Sometimes brewers add a second yeast later in fermentation to help finish the process after they've achieved the flavor profile they want from the primary yeast. This technique is popular with saison brewers who use the classic Dupont strain, which tastes great but poops out long before fermentation finishes. As a result, brewers will blend a second strain known for high attenuation with it. Several yeast companies offer just such a blend, or you can blend your own.

3. Bottling only: Using a different yeast for bottle conditioning is pretty common. A brewer might want a yeast that compacts on the bottom of the bottle or one that has a lower alcohol resistance to prevent it from eating complex sugars still in the beer and over-carbonating the bottles. For example, many German brewers filter out the yeast in hefeweizens and replace it with a lager yeast since the hefe yeast tends to autolyze (break down) much more quickly than lager yeast. Adding a different yeast at bottling for flavor development isn't going to do much, however, because the yeast won't undergo enough growth to produce any substantial esters. The one exception is adding *Brettanomyces* (page 130) at bottling, because it will produce a lot of funky aromas and tastes and will eat sugars that normal beer yeast can't.

American Ale

The most popular ale yeast strain among homebrewers and commercial breweries originated from the Sierra Nevada Brewery. (Rumor has it that they propagated it from the old Ballantine brewery.) It's sold as California Ale WLP001 by White Labs, as American Ale WY1056 by Wyeast, and as Safale US-05 in dry yeast form. Of the breweries in my city, three-quarters use Cal Ale as their primary strain!

Why is it so popular? Cal Ale has a clean flavor that lets the malt and hops shine through. No fruity or bready flavors muddy the waters. It doesn't flocculate too quickly, which means it does a good job of attenuating most beers and cleaning up any off-flavors created in fermentation. It's also a workhorse strain that can be repitched for 10 to 12 generations. That so many brewers use Cal Ale is a double-edged sword: It allows breweries to borrow yeast in an emergency, but it also gives many breweries the same house flavor (especially if they use the same base malt as well).

Many other American strains have more character than Cal Ale. California Ale V WLP051/American Ale II WY1272 is much fruitier and leaves a maltier beer. (If you have the choice, go for the WY1272.) Dry English Ale WLP007/British Ale II WY1335 also is characterized as a clean American strain because it has low fruitiness and good attenuation. (Stone Brewing Company supposedly uses it.) A seasonal offering from Wyeast, Pacman WY1764 is a popular yeast sourced from Rogue Brewing. It's a strong fermenter and good flocculator.

British Ale

Traditional British yeast strains are fruitier than Cal Ale, often with some bready notes. They're also known for flocculating more quickly than American strains. Since this will produce clearer beer in a shorter amount of time, it's a natural for beers served on cask. High flocculation also means the attenuation will be lower in most cases, however, and you may have to rouse the yeast or repitch to get the yeast to continue its work.

Dozens of British ale strains exist, and each can be quite distinctive. The most popular is English Ale WLP002/London ESB Ale WY1968, which comes from the Fuller's brewery. Like Cal Ale, it's another workhorse strain that ferments strong and fast. It leaves a mildly fruity, almost biscuity aroma and clears amazingly fast. I've had crystal-clear beer flowing from a keg one week after pitching the yeast! Other fairly popular British strains include London Ale WLP013/WY1028 (crisp, complex, almost woody) and Burton Ale WLP023/Thames Valley Ale WY1275 (very fruity and complex but buttery if fermented too warm). Irish Ale WLP004/WY1084 is smooth with a hint of butterscotch, a good yeast for high-gravity beers. One of my favorites is London Ale III WY1318. It clears quickly, contributes soft fruity notes, and is very easy to work with. When using any British ale strain, conduct a diacetyl rest at the end of fermentation (page 102).

Belgian Ale

It's impossible to say much that applies to all Belgian yeasts because each strain is so unique. Some are highly phenolic (bubblegum and banana notes), some are estery (pear and pineapple notes), and others are earthy with hints of dark fruit. To complicate further, most strains produce a different character when fermented at different temperatures! The same strain fermented at 65°F and 75°F produces two very different beers.

The modern method of fermenting a Belgian ale is to pitch the yeast at normal temperatures (65°F) and let it slowly rise over the course of the fermentation until it gets to around 75°F. (See page 115 for more information.) Warming it at the end of fermentation ensures that the yeast will keep eating any leftover sugars and reabsorbing any off-flavors. This is especially important when making strong Belgian ales, because it encourages a dry, crisp finish to the beer even when it rises above 8% ABV.

When I choose yeast for my Belgian ales, I think about the color of the beer. In light-colored brews, there isn't much malt flavor, so it's an opportunity to let a brighter, spicier yeast be the main focus of the beer. Some of my favorites are Belgian Ale WLP550/Belgian Ardennes WY3522 (reportedly from the Achouffe brewery) or any Belgian Wit strain. For amber to dark styles, find a strain that melds with the complex malt and dark candy-sugar flavors. Some of my favorites are yeasts that add a dark fruit and/or earthy element to the beer: Abbey Ale WLP530/Trappist High Gravity WY3787 (used by three of the top Trappist breweries), Abbey IV Ale WLP540/Belgian Abbey II WY1762, and Belgian Strong Ale WLP545 are all excellent. For a mild yeast flavor in a Belgian pale ale, try Antwerp Ale WLP515.

Saisons are delicious to drink, but they can be very problematic when it comes to fermentation. Many strains often stall out at 1.020 or higher. This can lead to bottle bombs and a sweet flavor unacceptable in a saison. Certain strains are more troublesome than others: Belgian Saison I WLP565/Belgian Saison WY3724 are sourced from Saison Dupont, which makes fantastic beer, but the yeast is very finicky. To help the fermentation go to completion, give it plenty of oxygen and drive the fermentation by increasing the temperature as it progresses. Pitch around 70°F and allow it to get up to 90°F by the end of fermentation if possible. The final gravity on saison strains can be amazingly low. Readings of 1.000 aren't uncommon. Packaging too early is a common mistake when brewing saisons. The gravity should be 1.006 or lower to avoid bottle bombs.

Many different saison strains are available, especially from new boutique yeast labs. My saison blend at Zebulon mixes Saison Blend #1 and Saison Blend #2 from the Yeast Bay in San Diego (along with some native yeast from Southyeast Labs in South Carolina). Remember that saison yeasts produce different flavor profiles at different temperatures. A strain might be boring at 65–68°F but amazing at 85–90°F.

Hefeweizens are unique in that the beers are almost defined by the yeast strains used. Descendants of a wild yeast, the strains can produce phenols that create a traditional clove aroma and taste. The vast majority of breweries use one specific yeast strain: It's available as Weihenstephan Weizen WY3068 from Wyeast and Hefeweizen Ale WLP300 from White Labs. They're both supposedly the same strain, but the two are slightly different. Do a side-by-side experiment with both strains. If I had to choose one, I'd go with the Weihenstephan Weizen WY3068. Like the other Belgian strains, these strains can produce very different beers at different fermentation temperatures. You can get too much bubblegum aroma and taste if you go too warm. Pitch at 65°F and ramp up to 68°F for the final few days. Some brewers will disagree with that method, but I think that it's best to bottle condition hefes. They should have a high level of carbonation, and it's very difficult to get that in a keg without lots of foaming.

Lager

There isn't much perceivable difference among lager strains. The low fermentation temperatures keep the yeast's fruity esters firmly in check, but subtle differences still exist. Some lager strains leave the beer a bit more malty, some a bit crisper, and a few are prone to leaving sulfur or buttery diacetyl behind. Lagers are enough work on their own—building up a big starter for proper pitching, lots of oxygenation, and maintaining proper fermentation temperatures. Why choose any strain that could prove problematic? With that in mind, avoid Southern German Lager WLP838/Munich Lager WY2308, which produces a lot of diacetyl, and Czech Pils WY2278, which can produce lots of sulfur.

Instead, pick one yeast strain that's easy to work with and familiarize yourself with it. Keep repitching it for a few batches. A great workhorse lager yeast and the most used widely lager yeast in Germany is German Lager WLP830/Bohemian Lager WY2124. If you're looking for more complexity, try the German Bock Lager WLP833 (from Ayinger Brewery). My favorite lager yeast is the Hessian Pils from the Yeast Bay in San Diego. It clears amazingly well without filtration and has a classic lager profile.

USING YEAST

One of the most common bad habits of new brewers is pitching just one vial or smack pack of yeast per 5-gallon batch. If a company handed you a couple of ounces of hops, is that all you'd use in your IPA? Just as with hops and malt, the amount of yeast required for each beer is unique. For the average homebrewed beer, a single vial or smack pack is barely at—and often far below—the bare minimum amount of yeast a professional brewery would use.

In addition to quantity, we'll also get into yeast health in this section. Yeast cells take in all necessary nutrients and expel all waste products through their cell walls. If they're not healthy, things go downhill quickly. You have to give your yeast what it needs. Remember, we aren't brewers, we're yeast farmers!

CELL COUNTS AND PITCH RATES

Pitching the proper amount of yeast is easy if you follow a few basic guidelines. On the following pages, we'll demystify cell count, learn a couple of ways to make starters, and cover plenty of other topics that will guarantee a happy home for your yeast—because happy yeast means happy beer!

A typical vial or smack pack contains somewhere in the neighborhood of 100 billion yeast cells. According to the manufacturers, that number is designed to be the proper amount to pitch directly into 5 gallons of wort at about 1.050 OG within a month of the manufacturing date. But the typical rate used by professional brewers is 1 million cells per milliliter per degree Plato (1.004), which equates to 4 billion cells per gallon for every increase of 1.004 of your gravity. Here is an example for 5 gallons of wort at 1.050 (which is 12.5 degrees Plato):

4 billion cells × 5 gallons × 12.5 Plato = 250 billion yeast cells

That means most professional brewers are pitching between double and triple the amount you get in one vial or pack! Does that mean that you really need to buy three packs? Yes, if you want to pitch the same amount of yeast a brewery would. But you have a few other options, as well.

➡ **Make a standard 2-liter starter (page 110):**
This roughly will double a colony of yeast, so you can take one vial or smack pack and turn it into two by making a mini-batch of beer. This amount, while still less than brewery levels, very rarely results in any subprime fermentations for 5 gallons of any ale up to about 1.060 (though I recommend doubling it by making two starters from two vials or smack packs if you're making a lager).

➡ **Brew a low-gravity beer to double as a starter:**
Whip up a delicious low-gravity beer, such as an ordinary bitter, when using a single vial or pack of liquid yeast for the first pitch. Make sure the recipe you choose is under 1.040, and leave as much sludgy hop residue in the kettle as possible so that it doesn't get mixed in with the yeast. One package of yeast never has a problem handling 5 gallons of wort at this strength. Then you can save the yeast from the bottom (or top, as on page 125) and use it to brew stronger beers. In essence, the beer you get to drink also acts as a 5-gallon starter! This is my preferred method when growing yeast for a strong ale (more than 1.070) or a lager.

➡ **Pick up fresh yeast slurry from a local brewery:**
This is by far the easiest way to get a healthy colony ready to pitch on the same day. Call ahead and bring a sanitized quart Mason jar. For a 5-gallon batch of 1.060 wort, you'll need only about 4 ounces (½ to ¾ cup) of yeast slurry if it's an ale and 8 ounces (1 cup) of slurry if it's a lager. Double-check to see what beer it's being pulled from, too. The yeast slurry from a strong IPA may contain too much hop character if you're brewing a drastically different style of beer.

Pitch Rates for Various Beer Styles

Beer Style	OG	Ideal Cell Count at Pitching	Number of Vials/ Packs without Starter	Starter Size with One Vial or Pack
Ordinary Bitter	1.035	100 billion	1	None needed
American Pale Ale	1.050	250 billion	2	2 liters
American IPA	1.065	300 billion	2	2 liters
Russian imperial stout	1.095	400 billion	N/A	Repitch onto a 5-gallon batch of low-gravity beer
Hefeweizen	1.050	150 billion	1	2 liters
Saison	1.055	250 billion	2	2 liters
Belgian Golden Strong	1.090	300 billion	3	Repitch onto a 5-gallon batch of low-gravity beer
Classic American pilsner	1.048	300 billion	3	1 gallon (with two vials or packs)
Dopplebock	1.070+	500 billion	N/A	Repitch onto 5-gallon batch of low-gravity beer

STARTERS

If money is no object, buy two or three packages of yeast to get the right amount for your beer. For the rest of us, starters are a great way to save money and begin fermentations with more yeast. A yeast starter is essentially a mini-batch of beer made so that your yeast culture can create more cells ahead of fermenting your actual beer. In addition to growing your cell count, a starter invigorates the yeast cells and gets them ready to start fermenting. Making a starter is an essential technique that any serious brewer should know how to do.

The traditional size of a starter is 2 liters (about $\frac{1}{2}$ gallon). You can make a starter about 24 hours before you brew and pitch it at the height of its activity, or you can let the starter finish fermenting, refrigerate it for two to three days, pour off the fermented wort, and pitch just the settled yeast. I prefer the latter method, because I don't want to dilute my beer with 2 liters of generic unhopped wort. However, in the rare instance that you

have a stuck fermentation and want to restart the fermentation with a starter, pitch the starter at the height of activity (after 12–24 hours). Also, if it's brew day and you forgot to make a starter, even a small starter made a few hours before brewing can be beneficial.

MAKE A STARTER

A 2-liter starter is the most popular size among homebrewers. It will double your yeast population in about 24 hours, so a vial or smack pack with 100 billion cells will become 200 billion cells. This is an acceptable, if not quite perfect, pitch of yeast for a 5-gallon batch of beer under 1.060. If you want to brew a bigger starter, consider brewing a 5-gallon batch of low-gravity beer. That way, you'll have some beer to drink instead of throwing out gallons of starter wort! Just remember: Regardless of the size or the beer you're pitching into, shoot for a gravity of 1.040 for the starter.

YOU NEED

1 medium to large pot

7 ounces light dry malt extract

2 liters purified water

Fermcap (optional)

sanitized container that holds 2 liters
   (a glass growler works)

sanitizer

aluminum foil

Note: Fermcap is the brand name of an anti-foam additive. It's optional but highly recommended. A drop or two in your starter will keep the yeast from foaming over the top. It also can help when boiling the starter, which is notorious for boiling over.

1. Boil 7 ounces of dry malt extract with 2 liters of purified water for 20 minutes. If using Fermcap, add 2 drops before the wort comes to a boil.

2. Pour this mixture into the sanitized container. Chill to the pitching temperature of your yeast. If using a glass growler, chill the wort below 120°F before pouring into the growler to prevent cracking.

3. Aerate by shaking for 45 seconds. Pitch your yeast into the starter and let it ferment for at least 24 hours. If you can shake the starter occasionally (every hour or so), you will get more yeast growth. The shaking releases CO_2 from the wort and introduces oxygen while keeping the yeast in solution.

4. After 24 hours, you can pitch the entire starter into your wort, assuming it's visibly very active. Alternatively, you can wait until the starter's activity dies down (typically 48 hours) and then refrigerate your starter. If you refrigerate your starter, take it from the fridge on brew day just as you would a vial or smack pack. While the starter is still cold, carefully pour off as much of the beer on top as possible without discarding any yeast. When your wort chills to pitching temperature, shake the yeast into solution and pitch into your wort or add some of the fresh wort to the yeast sediment and swirl it around. Let it reactivate the yeast for 30–60 minutes, and then add it to the wort—this will wake the yeast cells and ready them for action.

Note: If you brew frequently, you may want to make a large batch of wort and package it in large Mason jars so you'll have a supply of it on hand. If you heat the sealed jars in a pressure cooker, they can be stored at room temperature. However, if they're just filled with boiling wort, then you need to store them in the fridge.

VIABILITY

You need healthy yeast to have a healthy fermentation. Homebrew stores tend to receive refrigerated shipments, which means the yeast has stayed in ideal storage conditions since being made. Yeast shipped straight to the consumer without refrigeration will fare worse. Even under ideal transportation and storage, White Labs estimates that yeast viability drops to 75 to 85 percent after 30 days. If your yeast is older than that or shipped warm, definitely make a starter to get back to the original amount of cells. Generally, if you buy yeast from the homebrew store, assume that every 30 days the cell count drops by 20 percent. If you have a yeast package that's three months old, it's sitting at 40 percent of 100 billion cells.

Variation: Advanced Starter

If you're looking for maximum cell growth in a starter, then you need a stir plate. A constantly agitated and aerated starter increases cell counts by two or three times more than a standard starter.

YOU NEED

equipment and ingredients listed on page 111, except the sanitized container

stir plate

stir bar

2-liter Erlenmeyer flask (which can be boiled on a stovetop and then chilled without cracking)

1. Combine 7 ounces of dry malt extract, your stir bar, and 2 liters of purified water in the Erlenmeyer flask. **A**

2. Add Fermcap if desired and bring to a boil. **B** Boil for 20 minutes. **C**

3. Chill to the pitching temperature of your yeast.

4. Pitch your yeast, place your flask on the stir plate, and turn it on. **D** The stir bar will start to spin inside the flask, creating a whirlpool. Fermentation should finish within 24 hours.

5. After 24 hours, you can pitch the entire starter into your wort, assuming it's visibly very active. Alternatively, you can wait until the starter's activity dies down (typically 48 hours) and then refrigerate your starter. If you refrigerate your starter, take it from the fridge on brew day just as you would a vial or smack pack. While the starter is still cold, carefully pour off as much of the starter wort as possible without discarding any yeast. When your wort chills to pitching temperature, shake the yeast into solution and pitch it into your wort.

Variation: Base Beer as a Starter

To make world-class lagers, it's essential to have a large quantity of healthy yeast. It's difficult to get enough yeast by making a starter, so I recommend making a "sacrificial" batch. This batch should be drinkable, but making it is really about creating a 5-gallon starter so you can harvest the yeast. You also can make a starter for a high-gravity ale with the same process.

YOU NEED

basic brewing equipment (page 3)

For a lager starter: 7 ½ pounds pilsner malt, 1 ounce low-alpha hops (4% alpha acid), and 1 pack or vial yeast

For a high-gravity ale starter: 7 pounds pale-ale malt, ½ pound crystal 60°L, 1 ounce low-alpha hops (4% alpha acid), and 1 pack or vial yeast

1. Brew your beer as usual. Make sure that your gravity is about 1.036–1.040 and that you have just enough hops for balance (generally 15–20 IBUs).

2. Oxygenate well before pitching your yeast. (See page 116.) Ferment the beer at room temperature (65–70°F).

3. When this batch finishes, you will have enough yeast at the bottom of the fermentor to fill up a quart-size Mason jar. This is the proper amount of yeast to pitch for a lager or an ale over 1.080. (It's also enough yeast to pitch into two or three batches of a normal gravity ale.) In general, you need ½ to ¾ cup of yeast slurry for ales and 1 to 1 ½ cups for lagers. Store the Mason jars in the fridge for no more than three to four days before you brew your "real" lager or ale.

Note: If you want to go further into lab techniques, such as growing yeast on plates or in test tubes, pick up a copy of *Yeast: The Practical Guide to Beer Fermentation* by Chris White and Jamil Zainasheff.

USING DRY YEAST

Back in the early days of homebrewing, dry yeast was the only option, and it was terrible. Homebrewers often had nothing better to use than the repackaged bread yeast found under the cap of a can of malt extract! Today, dry yeast has evolved into a respectable option. Many small commercial breweries use it because it's cost effective and can be stored easily. There's an advantage to dry yeast for homebrewers, too. Danstar and Fermentis, the two largest dry-yeast companies, both package their yeast in an 11-gram foil pack that contains 200–400 billion cells. As you might remember from page 109, that's the perfect size for most normal 5-gallon batches.

The main downfall of dry yeast is lack of selection, although prospects are improving slowly. One of the most popular strains of dry yeast is Safale US-05 (close to the same strain as the one used by Sierra Nevada), which is crisp and clean with soft fruit notes. Another is Safale S-04, a British strain that clears faster than the US-05 but gives beer a breadier, fruitier character. Both are available at any homebrew shop. There also are new dried yeasts for brewing lagers, hefeweizens, wit beers, Belgian ales, and saisons, but reviews have been mixed, and the prices can be similar to liquid yeast.

Dry yeast needs to be rehydrated properly before pitching. Boil ½ cup of purified water and pour it into a sanitized cup or jar. Let it cool to 95–105°F, sprinkle the dried yeast on the surface of the water and leave it for 10 minutes. Stir the yeast into a slurry with a sanitized utensil, and then pitch it into your wort. Oxygen isn't as critical with dry yeast because the manufacturers pre-oxygenate the yeast before drying. Shaking your fermenter for a minute should be fine.

Don't risk trying to repitch dry yeast. Rehydrate a fresh pack for every beer. There's always a tiny amount of contamination in dry yeast that you won't notice in one generation, but, after multiple uses, it may start to show up. It's not worth the couple of bucks you'll pay for a fresh pack.

YEAST HEALTH

Pitching the right amount of yeast is only half of the battle. You also need to create an environment that will let it happily turn your wort into beer. The following details separate good brewers from great brewers. Without paying close attention to fermentation temperatures and proper aeration/oxygenation, it will prove almost impossible for you to make consistently great beers.

Fermentation Temperature

Aside from sanitation and pitch rate, the most important point to focus on is fermentation temperature. If you've ever had a beer that smelled boozy, tasted overly fruity, or gave you a headache after drinking it, it was probably due to a hot fermentation. Furthermore, you can't reproduce a beer consistently unless you can control the temperature within a few degrees. Let's start with the two most common problem areas, since they're responsible for the majority of yeasty off-flavors.

➡ **Temperature at the pitch:** While yeast companies recommend pitching yeast on the warm side and then cooling the wort to the proper temperature once fermentation has started, it's better to do the opposite. Chill your wort (overnight if necessary) to the proper pitching temperature before aerating and pitching the yeast. This one method will take your beers, especially lagers, to a new high.

The science behind this advice is that the yeast produces most of the precursors for potential off-flavors during the growth phase, which happens in the first 48 hours. If you pitch at the proper temperature and keep the beer cool for the first two days of fermentation, you can relax a bit. Many breweries purposely let the beer warm up as the fermentation progresses, a process known as "driving the fermentation."

For lagers, it's doubly important to cool the wort to pitching temperature before adding the yeast. When I first started brewing, it took me years to realize that adding the yeast at 68°F and then putting it into a chest freezer set at 50°F wasn't achieving the clean flavor I wanted. When I waited overnight, letting the wort drop to 47°F, before adding oxygen and the yeast, my lagers took a big jump in quality.

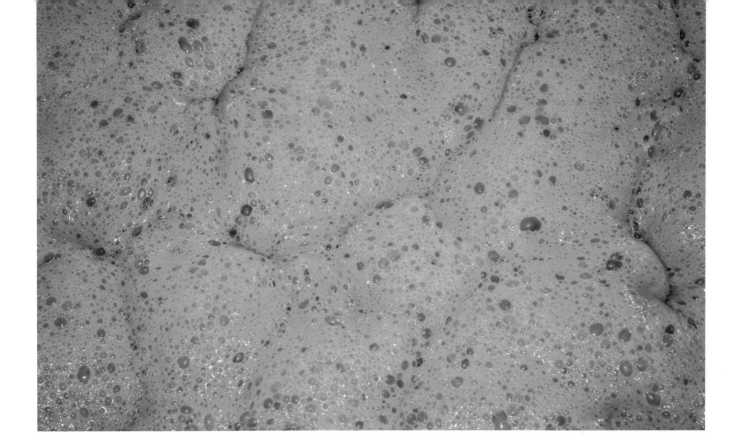

➡️ **Beer temperature vs. ambient temperature:** Many new homebrewers set their beer in a room that's the temperature they want for their fermentation and assume everything's fine. But beer inside the fermentor can measure as much as 10°F more than the ambient temperature, which means that, even if you set your beer in a 70°F room, it can heat to 80°F during the peak of fermentation. That's a huge difference! To get an accurate reading of the beer, you should insert a thermometer in the beer itself. Since that's not practical for sanitation reasons, do the next best thing: Tape the thermometer to the outside of the fermentor and insulate it. A piece of polystyrene foam or a similar product works well. Inexpensive adhesive strip thermometers also work fine, but you'll need to replace them occasionally.

During the first few days of fermentation, when the yeast is churning up the most heat, it's essential to monitor and control the temperature. If it gets too hot and you don't have a fermentation chamber with temperature control, a cold-water bath can help. Once you're out of the danger zone, the temperature in the bucket or carboy will settle out much closer to your ambient temperature.

Fermentation Schedules

Pitching the yeast cool and starting the fermentation cool for the first couple of days before letting it warm up near the end is known in commercial brewing as "driving the fermentation." This practice minimizes the production of off-flavors and makes cleaning them up easier for the yeast down the line. After three to four days, most fermentations are half finished at least, and the yeast cells are having a midlife crisis—they're getting tired and don't particularly enjoy the rising alcohol content. This is the time to let the temperature start rising a few degrees. It will keep the yeast cells active and help them work on that last little bit of fermentable sugar. Warming up the fermentation toward the end also increases the yeast's ability to reabsorb diacetyl and other off-flavors.

For most beers, you want a gentle increase in temperature: around 5°F for an ale and 10°F for a lager or a Belgian ale. After four to five days at the warmer temperature, the beer should be completely fermented out. You'll be able to tell: There will be no activity in the airlock and no krauesen (foam) on top of the beer. Hydrometer readings should show the same final gravity consistently. Every beer should stay in the primary fermentor for at least two weeks to ensure complete fermentation and diacetyl cleanup. Lagers or beers fermented cool (Kölsch, alt, cream ale) should go for three weeks before transferring off the yeast or chilling.

You can try to drive fermentation temperatures based on cool or warm areas of your house, but to control temperatures effectively you'll need a chest freezer or fridge with a temperature regulator eventually.

American and British Ale Fermentation

1. Pitch the yeast at 65°F and hold the temperature for 3 days.

2. Let the temperature rise slowly to 68°F over the next 4 to 5 days.

3. Let the temperature rise to 72°F over the next 3 to 5 days.

4. At this point, commercial breweries crash the beer to 32°F to flocculate the yeast from solution. You can do the same using a refrigerator or chest freezer with a thermostat. If you're kegging your beer, this is the time to add a fining agent if you're using one (page 123). After 3 days of cold conditioning, transfer the beer into a keg or bottling bucket.

Traditional Belgian Ale

1. Pitch the yeast at 68°F and let it rise slowly to 76–86°F over the course of 7 to 10 days.

2. When the fermentation finishes, let the beer sit at room temperature for several weeks.

3. Bottle condition with fresh yeast.

Modern Belgian Ale

1. Pitch the yeast at 66°F and let it rise slowly over the next 7 to 10 days to 68–70°F.

2. Raise the temperature to 72°F and keep it there for 3 to 4 days.

3. When fermentation finishes, drop the temperature to below 40°F and hold for one week.

4. Bottle condition with fresh yeast.

Traditional Lager

Historically, lager fermentation took place in an ice-cooled room. Once the yeast started to slow down and stopped producing heat, the temperature started to drop. Instead of increasing the temperature at the end of fermentation as brewers do today, the opposite happened, making it much harder for the yeast to reabsorb diacetyl and other off-flavors. This meant the beer had to sit for months at 32–36°F in order to clean up. If you want your lager finished faster, go with the Modern Lager schedule below.

1. Pitch the yeast at 46°F and hold the temperature for 7 to 10 days.

2. Slowly drop the temperature of the beer 1–2°F a day until it reaches 38°F.

3. Store the beer at 34–38°F for 3 to 12 months.

Modern Lager

1. Pitch the yeast at 46–48°F and hold the temperature for 4 to 5 days.

2. Slowly let the temperature rise to 55°F over the next 5 days.

3. Raise the temperature to 68–70°F for 5 to 7 days for diacetyl reduction.

4. When fermentation finishes, drop the temperature to 32°F and hold for a week before transferring into a keg or bottling.

Aeration/Oxygenation

Yeast cells can reproduce anaerobically (without oxygen), but the daughter cells will be weak, which often leads to off-flavors and poor fermentations—especially if you plan to reuse the yeast. For healthy yeast, oxygenation is key.

The optimal amount of oxygen fluctuates based on the yeast strain, the gravity of the beer, and many other variables. But the rule of thumb is 10 ppm for ales and up to 15 ppm for lagers, strong beers, and saison strains. Proper oxygen levels are even more important if you're planning to reuse your yeast. Oxygen-deprived yeast can start acting badly in a couple of generations. You can introduce oxygen in a few ways.

Once you buy the equipment and fill the tank, adjust the regulator so that the flow is 1 liter per minute. Most beers require just one minute of flow at that rate, which will give you around 10 ppm. Lagers and stronger beers will require two minutes, which should give you around 15 ppm. Over-oxygenation isn't really a concern unless you accidentally set it and walk away.

When using pure oxygen, oxygenate before you add the yeast. There's a chance that the pure oxygen (O_2) may freak the yeast out. Don't wait too long to add the yeast after oxygenating, though, as it starts coming from solution almost immediately. Once you add the yeast, it will absorb all the oxygen from the wort within an hour or two, so you may want to add an additional charge of oxygen after 12 hours for high-gravity beers.

➡ **Aquarium pump setup:** Some homebrewers use an aquarium pump fitted with a diffuser and a sterile in-line filter. These pumps use air and not pure oxygen, so you're still limited to 8 ppm. Go for an O_2 tank/regulator/diffusion stone if you have the money to spare. If not, you might as well just shake your fermentor vigorously, which will get you in the 8 ppm range for free.

➡ **Manual labor:** Splashing your wort into the fermentor post-boil using a siphon spray will give you only around 4 ppm. Pouring wort back and forth between buckets greatly increases your chances of an infection and isn't recommended. Shaking is the best way to oxygenate without additional equipment. Shaking for just 45 seconds will oxygenate your wort as much as possible. There's no benefit to shaking for 10 to 20 minutes. Since you're not using pure oxygen, your max is about 8 ppm no matter how long you shake. Once you're done shaking, pitch the yeast immediately.

➡ **Oxygen tank and diffuser:** The best and most expensive option is a tank of pure oxygen with a diffusion stone and a regulator. The oxygen tank is the pricey part. You can get one through a homebrew supply store or through a welding or CO_2 supplier. Look for oxygen regulators on eBay—they're remarkably cheap. Just make sure you find one that allows you to adjust the flow of the gas. Most online homebrew suppliers sell diffusion stones. I like the kind attached to a stainless wand.

Yeast Nutrients

For the most part, yeast has enough nutrients in a typical all-malt beer, so you don't need additional nutrients. But yeast nutrients are like multivitamins: they may not do much, but why not use them? I typically use $\frac{1}{2}$ teaspoon of Wyeast nutrients in a 5-gallon batch. On the other hand, I've never noticed any yeast health problems when not using nutrients, even for 10-plus generations.

Nutrients supply some nitrogen, some B vitamins, and often some zinc. (There's some controversy about whether an all-malt wort has enough zinc for normal yeast health, so it can't hurt to get a nutrient that includes it.) A brand of nutrient sold as Servomyces consists of dried yeast fed a large amount of zinc. This allows German brewers who can't use nutrients if they follow Reinheitsgebot (page 157) to add zinc. Yeast cells also need at least 50 ppm of calcium and 10 to 20 ppm of magnesium, which is usually present in the water supply. However, if you're using reverse-osmosis or very soft water, adding $\frac{1}{2}$ teaspoon each of Epsom salts and gypsum per 5 gallons is a good idea. (For more on water adjustment, see page 56.)

BELGIAN BLONDE ALE

While I enjoy all the beer styles of Belgium, some of them are too boozy to have on tap for regular drinking. A Belgian blonde is an exception. It has the same ABV as a pale ale but with the complex yeast character that makes Belgian beers distinctive. This is a good recipe if you want to try different Belgian yeasts and become accustomed to the fermentation regimen used for them. Belgian brewers commonly blend pilsner malt and pale-ale malt to add complexity. This recipe follows their lead and also adds a small amount of honey malt and aromatic malt for some additional complexity that doesn't overly darken the beer.

YOU NEED

basic brewing equipment (page 3)

9 gallons filtered water (page 12)

5 pounds German pilsner malt (46.2%)

4 pounds British or Belgian pale-ale malt (36.9%)

1 pound wheat malt (9.2%)

½ pound Belgian aromatic malt (4.6%)

⅓ pound honey malt (3%)

7.5 alpha acid units German Perle or Hallertau hops at 60 minutes (26 IBU)

1 Whirlfloc tablet

2 vials or packages the Belgian ale yeast your choice (or a 2-liter starter made from 1 pack, page 110); I used Belgian Ardennes WY3522

5 ounces dextrose/corn sugar (optional, use only for bottling)

TARGETS
Yield: 5 gallons
OG: 1.057–1.059
FG: 1.010
IBU: 26

1. Mix the malt with 4 gallons of water at 163°F or the appropriate temperature to mash at 148°F. Mash for 60 minutes. If you can, raise the temperature to 164°F for a mashout (to ensure complete gelatinization). If you can't mash out, don't sweat it.

2. Recirculate the wort until it's fairly clear. Run off the wort into the kettle.

3. Sparge with 5 more gallons of water at 165°F. Run off the wort into the kettle.

4. Bring the wort to a boil. Boil it for 30 minutes. Add the hops and continue to boil for 60 minutes. Add the Whirlfloc tablet at 30 minutes. Put your wort chiller into the wort at least 15 minutes before the end of the boil.

5. When the boil finishes, cover the pot with a lid or a new trash bag and chill to 65°F. Siphon the wort into your sanitized fermentor and pitch two packs of liquid yeast or a 2-liter starter.

6. Ferment at 65°F for two days, then allow the temperature to rise to 74°F and hold it there for two weeks.

7. Keg or bottle the beer. (If you're bottling, I recommend 5 ounces of dextrose/corn sugar for this beer.)

NEW AMERICAN PILSNER

I love sweet pilsner malt balanced with a firm hop bitterness and a clean,
flowery aroma. I had good luck a few years back brewing a pilsner using
100 percent Centennial hops, so I decided to see if I could push it even further
into the American hop realm. Because it includes dank, citrusy American
hops, this recipe combines the aggressive hop character of an IPA
with the drinkability and finesse of a German pilsner.

YOU NEED

basic brewing equipment (page 3)

8 gallons filtered brewing water (page 12)

9 pounds German pilsner malt (100%)

6.3 alpha acid units Citra hops at 90 minutes
(20 IBU)

3.2 alpha acid units Columbus hops at
30 minutes (7 IBU)

15 alpha acid units Columbus hops at
20 minutes (15 IBU)

6 alpha acid units Cascade hops at 10 minutes
(7 IBU)

½ ounce Citra hops at end of boil

1 Whirlfloc tablet

3 vials or packages German Lager WLP 838/
Bavarian Lager WY2206 yeast (or a 1-gallon
starter made from 1 pack; page 110)

½ ounce each Citra and Centennial hops (dry hop)

4.6 ounces dextrose/corn sugar
(optional, use only for bottling)

TARGETS

Yield: 5 gallons

OG: 1.050–1.053

FG: 1.010

IBU: 49

1. Mix the malt with 3 gallons of water at 165°F or the appropriate temperature to mash at 150°F. Mash for 60 minutes.

2. Recirculate the wort until it's fairly clear. Run off the wort into the kettle.

3. Sparge with 5 more gallons of water at 165°F. Run off the wort into the kettle.

4. Bring the wort to a boil. Add the first addition of hops and continue to boil for 90 minutes, adding the other hop additions as called for at left. Add the Whirlfloc tablet at 30 minutes. Put your wort chiller into the wort at least 15 minutes before the end of the boil.

5. When the boil finishes, cover the pot with a lid or a new trash bag and chill it to 48°F. If you can't get the wort down to that temperature, siphon it into your sanitized fermentor and chill it overnight so that it reaches 48°F. Then pitch three packs of liquid yeast or a 1-gallon starter.

6. Ferment at 48°F for 2 days, then raise the temperature to 50°F for 3 more days. Up it to 55°F for another week, add the dry hops, then let the temperature rise to 65–68°F for 3 days for a diacetyl rest and to complete fermentation. Crash to 32°F and hold for 7–10 days.

7. Keg or bottle the beer. (If you're bottling, I recommend 4.6 ounces of dextrose/corn sugar for this beer.)

YEAST AFTER
FERMENTATION

FILTERING AND FINING

You might think a finished fermentation means the yeast is done, but that's not the case. Even after you reach final gravity, yeast continues its work. If you bottle, the yeast will wake up and work hard one last time to carbonate your beer.

But let's start where we just left off: finished fermentation. When that happens, some yeast cells die. These cells settle to the bottom of the fermentor, which is why brewers throw away the very bottom of the yeast sediment cake—it's mostly dead yeast and other waste (proteins, hop particles, etc.) known as trub. If dead yeast remains in the beer, it eventually will autolyze, or leak its insides into the beer. This process results in a meaty, brothlike flavor that's unpleasant and a major flaw in any beer.

Healthy yeast doesn't start to autolyze for a long time, even after it dies. Homebrewers used to believe that beer had to be taken off the yeast after one week or it would develop off-flavors, but we know now that this isn't true. There shouldn't be any off-flavors from dead yeast until at least four weeks in the primary fermentation vessel—maybe up to six weeks if it's kept cold. That's one reason that most of today's homebrewers don't bother with secondary fermentation.

The other reason is that removing the beer from all of the yeast too soon (as soon as fermentation finishes) means that only some of the yeast—whatever was left in suspension—has a chance to reabsorb diacetyl and other off-flavors. If there's not enough yeast to do the job, those off-flavors stay in the beer. Leave just about any beer with all of its yeast for a total of two weeks before racking or packaging, just to be safe.

Having filtered beer through a variety of filters over the years, I don't recommend it on a homebrew scale. You're increasing the chance of oxidizing or contaminating your beer when you can get the same results with finings and a few days of patience.

On the finings side, your best option is gelatin. It's easy to work with, and it performs very well. The only downside is that it's not suitable for vegetarian or vegan consumers. A product called Biofine is vegan and works similarly. Isinglass, often mentioned in older brewing literature, isn't vegetarian. It's notoriously hard to work with and difficult to find. Since there's no advantage to using it over gelatin, just stick with gelatin.

No matter what you choose as a fining agent, the process is more or less the same. Here are the two types of fining I recommend:

➡ **Pre-fermentation:** Kettle finings coagulate many types of proteins in the boil, which means a clearer wort going to the fermentor. Almost all kettle finings derive from a seaweed called "Irish moss." You can find straight Irish moss, but use either the powdered version or a tablet version. (The most popular brand is Whirlfloc.) No matter what you buy, the cost is very small for each 5-gallon batch. Instructions for that amount usually appear on the package, but all of the various forms are added toward the end (final 15 minutes) of the boil.

➡ **Post-fermentation:** Gelatin typically gives you a beer that looks filtered within a week. Once your beer finishes fermenting and cleaning up (typically around two weeks), crash it down to 32°F overnight. This is when the beer develops the chill haze you're trying to avoid—clarifiers can't clear something that's not formed!

The next day, take a pack of unflavored gelatin (next to the flavored gelatin in most supermarkets) and slowly sprinkle it onto the surface of a cup of cold, purified water. After letting it rehydrate for 5 to 10 minutes, heat it to a boil in the microwave or on the stovetop and then immediately cut the heat. Add this to your chilled beer and give it a gentle swirl. Your beer should be crystal clear within a few days.

YEAST MANAGEMENT

When it comes to yeast handling and maintenance, most small brewpubs and breweries are counting their yeast cells in bucketfuls or seconds on a pump. Rarely is there a lab on the premises. But the estimation method does work. It may not be a textbook approach to cell counting, but as long as you don't repeat it for too many generations, the risk of a problematic fermentation remains very low.

The same goes for homebrewing. Everyone says that pitching onto a previous batch's yeast cake is taboo, but it works fine—as long as you do it only one time. If you forget to add oxygen to a batch, don't freak out. Just don't reuse that yeast. The most common problems that develop with yeast take multiple generations to show up, so, if you repitch only one or two times, you can relax a little. Since fresh yeast is inexpensive, start with a fresh culture after no more than four uses/generations.

Washing and Packaging Yeast

Commercial brewers often joke that they're really yeast farmers, because yeast in a brewery is rarely grown from a new culture but is harvested from the bottom of a fermentor and repitched in another batch instead. Some breweries go years without a new culture. As long as their sanitation is good and they treat the yeast properly, there's not much of a downside.

It's tempting for a homebrewer to do the same, continually saving yeast from one batch and using it in another. But the differences among commercial breweries and homebrewers are large in some important respects. Breweries brew every day, yeast goes from one fermentor and into the next in a matter of minutes, and rarely do breweries store yeast for weeks before using it. Commercial brewers are harvesting the yeast from fermentors with valves at the bottom, which allows them to discard the first stuff to come out (mostly hop residue, dead yeast, and break material) and get to the creamy white yeast in the center. Homebrewers, on the other hand, typically have a jar full of all kinds of junk that may contain only 10 percent healthy yeast! Trying to separate the healthy yeast from the detritus is a lot of work, but it isn't difficult, and it will result in cleaner and better yeast for your next fermentation.

YOU NEED

sanitizer, such as Star San or iodophor

2 (1-quart) Mason jars with lids (or other similar-size glass containers)

2 (1-pint) Mason jars with lids

beer that has finished fermenting (ideally one with minimal trub and hop residue mixed with the yeast and no more than 6% ABV) as well as any kegging or bottling equipment you need to package the beer

1 gallon pre-boiled and cooled water or commercial spring water

1. Mix your sanitizer and sanitize the glass containers and lids.

2. Siphon the beer into the keg or bottling bucket, leaving about an inch of beer at the bottom of the fermentor above the yeast sediment.

3. Swirl the fermentor to get the yeast into solution, then pour or siphon it into the quart jars until they're about two-thirds full. Top the jars off with the purified water, give them a good shake, and let them sit for about 15 minutes.

4. After 15 minutes, you should be able to see some separation in the jars. Carefully pour the top portion of the jars into the two sanitized pint jars, leaving behind the precipitate at the bottom of the quart jars. The pint jars can be stored in the fridge for up to two weeks; each is the proper pitching size for a 5-gallon batch of normal gravity beer.

Top Cropping Yeast

Back in the days before conical fermentors, yeast used to be collected from the top of an actively fermenting beer. This yeast is usually very healthy and clean because it's free of any dead cells or trub (which sink to the bottom), and the yeast is removed before the alcohol content of the beer gets very high. This is still the preferred way to harvest yeast for reuse, but it's tricky for commercial brewers because they don't have access to the top of the wort in a closed conical fermentor. Homebrewers have it easier if they're fermenting in buckets. They can crack the lid carefully and scrape off the yeast. Sanitation is of utmost importance when harvesting yeast this way. Enlist an assistant to hold the jar or freezer bag as you work.

YOU NEED

**wort 2–3 days into fermentation
   (too early, and the krauesen won't have
   formed properly; too late, and it might
   have fallen back into the beer already)**

sanitized stainless steel spoon

sanitized large-mouth pint jar or large freezer bags

1. In a draft-free room, carefully crack open the lid of the fermentor. Use the sanitized spoon to scoop off some of the yeast and deposit it quickly into the jar or freezer bag. Repeat until you have skimmed off most of the yeast. Quickly close the fermentor's lid and seal the jar or bag.

2. Keep the jar or bag in the fridge for up to a week. If you're using a freezer bag, place it in a second freezer bag to prevent it from contamination.

3. When you're ready to use the yeast, just add it to the wort. The full amount of yeast you collected is a good amount to pitch into a 5-gallon batch.

WILD BEERS *and* WOOD AGING

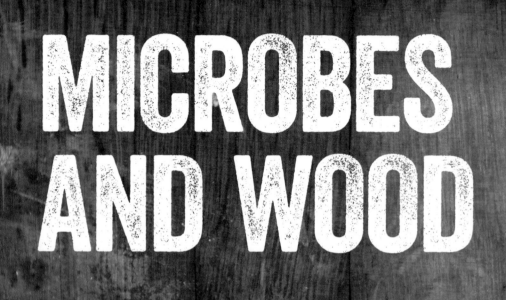

MICROBES
AND WOOD

A BRIEF INTRODUCTION

The use of different kinds of microbes instead of brewing yeast and the aging of beer in barrels may seem like a new frontier in brewing. But the use of wild yeast, bacteria, and barrels is hardly revolutionary. For centuries, brewers in a region of Belgium near Brussels added no ale or lager yeast at all. Instead, their wort cooled in large, shallow containers called "coolships," where it was exposed to the wild yeast and bacteria in the air of the region. The microflora took over the beer and started what's known as spontaneous fermentation. Brewers in this area found that these yeasts and the bacteria in the air made a delicious sour beer. In order to develop the desired flavors, the process often took several years. The wooden barrels in which the beer was stored contributed their own plethora of micro-organisms to the beer as well. To achieve a consistent product, brewers blended multiple barrels together.

Note: Sour beers from this region of Belgium are called "lambics" collectively, which can refer to traditional, take-the-enamel-off-your-teeth sour beers (Cantillon), or to sweet/sour fruit beers (Chapeau). The traditional Belgian lambic is called "gueuze" (pronounced GUH-za) and is a blend of one-, two-, and three-year-old lambics. If you want to taste a true example of spontaneous fermentation and the art of blending, look for beers from the Boon, Cantillon, or Drie Fonteinen breweries.

Beer traditionally was stored in barrels in other countries as well. But wooden barrels (usually made of oak) often were lined with a tarlike substance called "pitch" that prevented the beer from coming in contact with the wood itself, so the beer had little, if any, wood flavor. Even without pitch, the barrels usually were made of a tight-grained oak from Eastern Europe treated with steam to remove most of the wood flavor and kill most of the microorganisms living in the wood. Only Belgian brewers made a great effort not to get rid of the wild yeast and bacteria that lived in the barrels, just giving the barrels a quick rinse before refilling them.

Today brewers play with wild yeasts and bacteria (often called "bugs") in isolation and in tandem with wood. Some breweries take the more traditional approach with coolships and an extensive barrel program, while others add pure cultures of *Brettanomyces, Lactobacillus,* and other bugs, in modern stainless fermentors.

For homebrewers, both advantages and disadvantages stem from batch size and space restrictions. Sour beers take a long time to develop, and it's impractical for the average homebrewer to devote an entire wing of his or her house to the storage of multiple batches and barrels. However, homebrewers have an advantage when it comes to trying something new. A commercial brewery may shy away from using *Brettanomyces* and other assertive cultures, but homebrewers can play with much less risk. After all, if something infects a home system, it's just a matter of buying a new bucket and replacing a few other plastic pieces.

There are excellent resources out there when it comes to sour beer and the sour tradition—and more and more research in this science comes out every year. (Read *Wild Brews: Culture and Craftsmanship in the Belgian Tradition* by Jeff Sparrow, as well as Michael Tonsmeire's excellent book *American Sour Beers* and entertaining blog at themadfermentationist.com.) The goal of this chapter isn't breadth, depth, or even breaking new ground. Rather, it's all about taking some of the current knowledge and applying it practically to the average homebrew system. Now, let's get wild!

TIMELINE OF A SOUR BEER

Before we get into a microbe overview and the brewing itself, let's sprint through a typical open-air lambic beer fermentation.

With open-air fermentation, the first couple of weeks are full of activity. Of the many cultures that inoculate the wort, some of these are spoilage bacteria related to *E. coli.* Luckily, these are killed off quickly as the pH of the wort drops and alcohol is produced. No known pathogens can survive for long in beer, but to be safe, don't sample the wort of a spontaneous fermentation during the first few weeks.

After a week or so, wild species of the genus *Saccharomyces* take over the ferment and quickly start to produce alcohol. This continues for a month or two until the majority of the sugar is consumed. Lactic acid bacteria then take over and start producing the sourness that we equate with sour beers. The two main microbes are from the genera *Lactobacillus* (Lacto) and *Pediococcus* (Pedio). Lambic brewers use high rates of hopping with aged hops to inhibit Lacto production in order to give other bugs a chance to do their work before too much acidity inhibits it. Pedio often produces large amounts of buttery diacetyl and can cause the beer to turn ropy or gelatinous. (Belgian brewers refer to this as "the beer getting sick.") Luckily, it usually will go away on its own over the rest of the fermentation process.

After six months or so, a variety of wild yeasts known as *Brettanomyces* usually takes over the fermentation and starts to produce the classic horsey, barnyard, and cherry-pie character that's associated with lambics. This part of the fermentation will continue over the next year with occasional guest appearances by the yeasts *Candida*, *Pichia*, and *Hansenula*. By the time the beer is ready to drink, it has been home to a myriad of microorganisms that have produced hundreds of complex flavors. This is why lambic is such a strange and wonderful type of beer—and one that's hard to ferment the same way repeatedly!

MICROBES

During a traditional lambic fermentation, dozens of microorganisms are living in the beer. It's impossible and impractical to discuss them all. However, it's important to understand the major players—especially the cultures you'll likely buy to do this at home. So let's start with a short exploration of what exactly you're pitching when you're making sour beer.

Saccharomyces

As you probably know, *Saccharomyces* is the genus of yeast used for the majority of brewing and baking. The name means "sugar fungus," and its ability to consume sugar and excrete alcohol is the main reason we all love it. (For more on *Saccharomyces*, see Chapter 4.) Why is it in the chapter on wild beer? Well, even the wildest beers typically start with a *Saccharomyces* fermentation.

The choice of *Saccharomyces* yeast for your sour beer is up to you. The flavor contribution overall will be minimal. A common *Saccharomyces* strain found in traditional lambics is *Saccharomyces bayanus*, also known as Champagne yeast. Today most brewers just use a generic ale yeast. You can experiment with a less attenuative yeast such as English Ale WLP002/London ESB Ale WY1968, which theoretically should leave more sugars for the bugs to feed on.

Brettanomyces

The most popular of the funky bugs come from the genus *Brettanomyces* (just Brett, for short). The strains can give your beer a huge range of aromas and tastes, from pineapple to barnyard. Hundreds of Brett strains exist in the wild, but yeast labs offer only a small handful. If you decide to experiment with different strains, keep in mind that it can take up to a year for the flavors to develop and mature fully, so have patience.

Traditionally, Brett wasn't added to beer on purpose. It just existed in the wood of the barrels or in the air around the brewery. When adding Brett intentionally, you have a couple of options. It can be the only yeast you add to your beer, since it will convert sugars into alcohol just fine on its own for primary fermentation. You might think this would create a very Brett-y beer, but strangely enough Brett tends to produce a fairly clean-tasting beer when used as the primary yeast.

To coax stronger aromas from Brett, many brewers use it as they would for sour beer—pitching it after primary fermentation completes. You can do this in the fermentor when primary fermentation is about finished. You also can add Brett at bottling (a technique made famous by the Trappist brewery Orval), which will give you a slowly evolving aroma. You can get a slight tartness over time with Brett, but it alone won't make a sour beer.

The main concern with using *Brettanomyces* is the continuing slow fermentation as it consumes virtually everything. Many of my Brett beers have fermented down to 0.998 over the course of a year. This can be problematic when bottling because it will lead to over-carbonation and even bottle bombs. There's no real solution to this problem other than taking a hydrometer reading when you think the beer finishes, waiting three weeks, and taking another reading. If they're the same and below 1.005, then hopefully you're okay to bottle.

Since each Brett strain has its own characteristic flavors, let's run briefly through the four main varieties you can buy.

➡ **Brettanomyces claussenii (also known as B. anamolous):** This Brett originates on the skins of fruit and can be found in the air. It was found commonly in British beer aged in barrels as well. It often contributes a tartness and a slight earthiness, without the notable barnyard funk of other Brett strains.

➡ **Brettanomyces lambicus:** My favorite Brett of the bunch exhibits the classic cherry-pie flavor without the sweaty, barnyard notes of other Brett strains (notably *B. brux*, below).

➡ **Brettanomyces bruxellensis:** If you're looking for the classic, wild, funky Brett flavor that elicits strong reactions, *B. brux* is your strain. Think sweaty horse or an old barn, and you're getting close! That's not to say it tastes bad, though. The Belgian Trappist brewery Orval uses Brux as an addition at bottling, so grab a bottle and see if you fall in the love-it or hate-it camp. It can be cultured easily from a fresh bottle of Orval as well (page 142).

➡ **Brettanomyces bruxellensis trois:** Recently this strain has been shown not to be a *Brettanomyces* strain at all but a weird *Saccharomyces*. It produces some funk that you'd expect from a Brett, but technically it isn't Brett.

➡ **Other new Bretts:** Quite a bit of debate is taking place as to the taxonomy of the different Brett cultures, so expect a few name changes and new strains to hit the market over the next decade. The Yeast Bay and East Coast Yeast, both boutique yeast labs, specialize in new Brett strains and blends. If you're interested in this topic, subscribe to their mailing lists.

BRETT LOVES HOPS

The funky overripe fruit character that Brett adds to beer blends amazingly well with the citrus and tropical fruit aromas and tastes from modern hop varieties. Brett also will keep absorbing oxygen in the packaged beer, which keeps the hop aroma popping for months longer than a typical IPA.

Use a Brett blend such as Amalgamation from the Yeast Bay rather than a single strain, but, if you have to choose a strain, I recommend Brett brux. My hop recommendations include Citra, Galaxy, and any of the New Zealand varieties. Add the Brett after primary fermentation and wait a month or two until the FG is stable. Then dry hop heavily (4–6 ounces in a 5-gallon batch).

Lactobacillus

If you have had the pleasure of drinking a Berliner weisse or a gose, you've had a beer featuring lactic acid created by the *Lactobaccilus* bacterium. It's also what makes sauerkraut and pickles sour. Brewers use it as their primary souring bacteria because it can sour a beer in 24 hours, compared to months with *Pediococcus* (below.)

Brewers use a handful of Lacto strains, and your choice primarily comes down to whether you're going to do a kettle sour (page 143) or a mixed-culture fermentation (page 151). The most popular Lacto strain for quick kettle sours is *Lactobaccilus plantarum*. It's sourced easily from probiotic pills or drinks such as Goodbelly. The appeal of this strain is that it's inhibited by as little as 1 IBU of hops, which keeps it from infecting regular beer. It's also a homofermentative strain, which means it won't eat the sugar from the wort while it's souring. This is important if you'll be boiling the soured wort to kill the Lacto when it's done (more on that later).

Pediococcus

I don't know of any breweries that have produced a beer featuring *Pediococcus* alone. Pedio, like Lacto, creates lactic acid, but Pedio on its own can create major off-flavors in your wort, notably loads of buttery diacetyl. It also can give your beer the consistency of slime. Luckily, if you're co-pitching Brett along with the Pedio, the Brett will help clean up the diacetyl and slime, given enough time. The main appeal of Pedio is that it isn't hop sensitive like Lacto is. Belgian lambic brewers have been adding large amounts of aged hops to their sour beers for ages specifically because they want to inhibit the Lacto and give the other bugs a chance to grow. The slow-growing Pedio eventually will sour the lambic but only after everything else gets its turn at the table.

Since it's a team player, Pedio isn't as readily available from a lab or as easy to culture on its own as Lacto. But you never really want it working on its own anyway, so that's not a big deal. Just buy one of the commercial sour blends such as White Labs' Flemish Ale Blend WLP655, Wyeast's Roeselare Ale Blend WY3763, or culture some dregs from a brewery that has a nice sour program (such as Jolly Pumpkin). Chances are they have some active Pedio in their cultures.

Blended Cultures

Buying separate cultures of individual bugs can get expensive quickly, and the addition of specific individual cultures at specific times isn't well studied. So why not go with a blend?

The two most popular mixed cultures available for homebrewers are Wyeast's Roeselare Ale Blend WY3763 and White Labs' Belgian Sour Mix 1 WLP655, both of which are a proprietary blend of Brett, Lacto, Pedio, and some type of *Saccharomyces*. Other mixes exist and, as

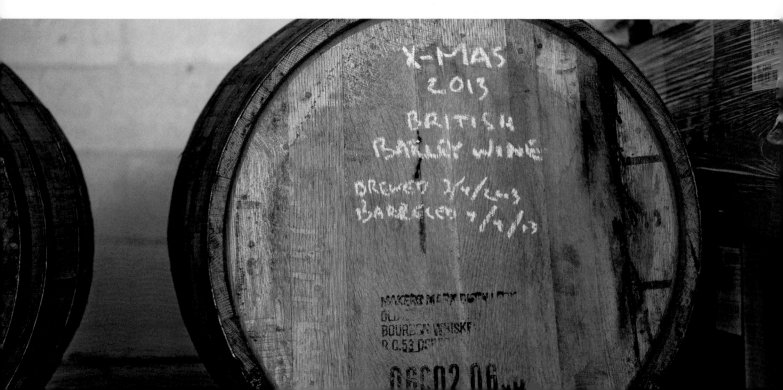

One more bug can turn beer sour—but not in a good way. *Acetobacter* creates a familiar flavor when it gets a hold of your beer: It turns the alcohol into acetic acid, which, as you may already know, is vinegar. It quickly can turn a batch of beer into a salad condiment, but it's fairly easy to keep it from thriving. *Acetobacter* needs oxygen, so, if your airlock is topped off and you don't keep cracking the lid to take samples, you should have no problem keeping it from your beer. If you want to add a hit of acetic acid to your beer (small amounts in Flanders red ales, for example), try adding a small amount of vinegar when you bottle.

long as brewers are interested, these labs will release more and more every year. Read their descriptions and try something new if it sounds like it will work for the type of beer you want to make.

Small yeast companies also have unique blends of mixed cultures that often are preferable to those from larger yeast companies. East Coast Yeast has a blend called BugFarm, and the Yeast Bay has a blend called Melange. These blends emulate traditional lambic cultures and take up to a year to complete fermentation. If you want a quick-turnaround sour beer, check out kettle souring (page 143).

Your final option is to follow the lead of Belgian brewers. When they find a barrel that's been working for them, they inoculate wort in that barrel and try to spread the culture mix to other barrels. How to do that without a barrel? Well, for homebrewers, the closest you can come is using cultures from breweries you like that make sours. One or two breweries pasteurize their sours (notably New Belgium), but most don't. That means you can buy beers from many Belgian brewers as well as domestic sour brewers, such as Jolly Pumpkin, and start your culture right from theirs. See page 140 for more on this method.

FORMS OF WOOD

Your choice of delivery method for wood flavor depends both on the type of beer and on how long you plan on aging it. If you plan to age a huge imperial stout for 6–8 months, then a spiral is perfect. But if you're looking for a hint of oak in a hoppy West Coast red ale, then medium-toast chips in the secondary fermentation with the dry hops are best.

Don't forget to sanitize any wood product before you use it. The preferred method is to place the product in a pot with a small amount of water and bring it to a boil. Steam it for 10 to 15 minutes, then wrap it in foil or put it in a plastic freezer bag until you're ready to use it. (For more on sanitizing a barrel, see below.) If you want a whiskey, rum, or other alcohol flavor in your beer, sanitize the chips, cubes, or spirals by soaking them in the alcohol of your choice for a few weeks instead.

Barrels

When you age beer in a used spirits barrel, the results can vary wildly. Barrel size, age, condition, and the original distillate all play a role. Barrel size is perhaps the most important variable. The difference in surface-area ratio of a 5-gallon barrel compared with a 60-gallon barrel is huge. The small barrel has about five times the ratio of the large barrel. For this reason, many brewers advise against aging any beer in a small barrel. The risk of rapid oxidation is just too large. However, if you do use a small 5- or 10-gallon barrel, the contact time needed to impart flavor could be as little as a few days or weeks.

When it comes to barrel age, new barrels aren't necessarily better. The first use of the barrel may yield over-the-top flavors in as little as a week, while the second fill may have to sit for months to achieve the same level of flavor. If you get your hands on a fresh, recently used spirits barrel, it should be self-sanitizing with 120- to 150-proof liquor oozing from its pores. If it's been sitting around for a few months, it needs to be filled with water for a few days so that the wood can swell up and seal it. Using really hot water (180–190°F) will kill any worrisome bugs as well. Whiskey barrels shouldn't have anything growing in them, but wine barrels most likely will, so soaking with very hot water is essential—unless the barrel was just emptied. Don't use any kind of sanitizer because it will soak into the wood, and you don't want that taste in your beer. Below you'll find a few more tips on working with barrels.

➡ The best advice I have on barrel-aging beer is to keep tasting. Take a sample after a week and then every week thereafter. When the beer has picked up enough flavor, rack it into a keg or bottles. (This is for clean beers only; sour beers aging in barrels should be tasted only every few months, if that.)

➡ Watch your oxygen pickup when using a barrel. The barrel naturally will let in small amounts of oxygen over time, so purge the barrel with a CO_2 tank, transfer without splashing, seal it with a good stopper and airlock, and check the airlock every couple of weeks to make sure it doesn't dry out.

➡ Once you have the beer from the barrel, you need to refill it quickly or it will turn vinegary on you. The optimal situation is to have another batch ready to go as soon as you rack from the barrel. If you can't manage that, fill the barrel with 180°F water, and bung it up. Use the barrel within a week, or cut it in half and use it as a planter.

➡ It's much easier for a commercial brewery that has beer around every day to manage a barrel program than it is for homebrewers. It can be an expensive and sometimes disappointing experiment, but that shouldn't stop you from trying if you're interested. It can be a fun big brew for a homebrew club as well.

Wood Pieces: Chips, Cubes, and Honeycomb

Chips are a more reliable, if slightly less impressive sounding, way to get wood flavor into your beer. The fragments go into the finished beer, just like dry hops, and start to contribute their taste in just a few weeks. In general, 1 ounce of chips in a 5-gallon batch gives a subtle woodsy note within two weeks; 2 ounces provide a more pronounced taste. Longer contact time will extract a little bit more taste, but most of it will be extracted from chips within two to three weeks.

Oak cubes are similar to chips but have less surface area, so it takes more time or more oak to get the same taste. Purists think that the thicker size means a variation of toasting throughout, which can add complexity. Honeycomb wood pieces theoretically fall somewhere between chips and cubes. Their main advantage is that they're available in many wood types besides oak.

Spirals

Oak spirals are long batons of wood that can be inserted into a keg or carboy. They aren't very popular with homebrewers at the moment because they can take a long time to contribute their oaky goodness. Expect at least a month of contact time to get a subtle taste, and up to six months for full extraction. Spirals are good candidates for big beers that will see extended aging, such as imperial stouts and barleywines.

TYPES OF WOOD

American, Hungarian, and French Oak

There isn't a huge difference in the various nationalities of oak, but I would choose French or Hungarian over American, which can have a sort of lumber-yard note to it. Hungarian oak is tight grained, which means it will take longer to release its taste. French oak is thought to be sweeter and more delicate. The more important attribute is the level of toast.

➡ Light toast provides a brighter, more fresh-cut log character than the darker toasts.

➡ Medium toast lends more warm, toasty aromas and tastes and is the most popular among brewers.

➡ Dark toast can give roasty, smoky notes but is designed mainly for aging liquors. It should be used with a very light hand in beers.

Other Woods

Oak has been used for centuries for storing beer, but there's no reason you can't use other woods to flavor your beer. You can find a wide variety of wood chips suitable for brewing in the barbecue section of a home-supply store. You also can find apple, cherry, mesquite, and hickory chips in the grilling section of some supermarkets. A few dollars will buy you more wood chips than you'll ever need in a lifetime of brewing.

For the tasting notes that follow, I put a handful of each wood variety in a 375°F oven for 30 minutes to sanitize them and add a nice warm, toasty character.

I then added 5 grams of each type of wood to separate half-gallon growlers of an ESB (the equivalent of almost 2 ounces in a 5-gallon batch). After two weeks of aging, I conducted a blind tasting among BJCP judges and fellow brewers. Here are the results.

➡ Mesquite had a tannic aftertaste and was very subtle. Some tasters swore that their barbecue synapses were being tickled. It would be safe to double the amount (4 ounces for 5 gallons).

➡ Apple also had a tannic though not unpleasant edge to it. There was a delicate sweetness and fruitiness. It might be nice in a saison, and it's a no-brainer for adding complexity to ciders.

➡ French oak was the ringer in the flight and was picked out immediately by all tasters. The toasty, woody aroma was stronger in this sample than it was in the others. It added a buttery, vanilla taste that blended nicely with the beer. The amount used seemed just right.

➡ Cherrywood was very subtle—almost undetectable.

➡ Hickory was the surprise favorite. It gave the beer a nutty, pecan-like aroma and taste that worked really well with an ESB. We added toasted hickory chips to several firkins of ESB at the brewery, and tasting-room customers liked it.

HICKORY-WOOD AGED BROWN ALE

Tyler Downey brewed this recipe over and over until he got it just right. It's an American-style brown ale with a decent amount of hops and alcohol. It has no crystal malt—just lots of toasty roasted malts that give it a complex backbone. We served this beer on cask with toasted hickory chips, but it's a good recipe for experimenting with the wood of your choice.

YOU NEED

basic brewing equipment (page 3)

9 gallons filtered brewing water (page 12)

10 pounds British pale-ale malt (90.5%)

6.4 ounces brown malt (3.6%)

6.4 ounces British chocolate malt (3.6%)

4 ounces Weyermann CaraAmber malt (2.3%)

Note: Despite the Cara in the name, this is actually more in the vein of a biscuit malt, which can be substituted

17 alpha acid units Columbus hops at 30 minutes (45 IBU)

1 Whirlfloc tablet

¾ ounce Columbus hops at end of boil

2 vials or packages California Ale WLP001/ American Ale WY1056 yeast (or a 2-liter starter made from 1 pack, page 110)

1 ounce hickory (or other wood), toasted in a 350°F oven for 15 minutes

3¾ ounces dextrose/corn sugar (optional, use only for bottling)

TARGETS
Yield: 5 gallons
OG: 1.056–1.058
FG: 1.011–1.013
IBU: 45

1. Mix the malt with 4 gallons of water at 169°F or the appropriate temperature to mash at 154°F. Mash for 60 minutes.

2. Recirculate the wort until it's fairly clear. Run off the wort into the kettle.

3. Sparge with 5 more gallons of water at 165°F. Run off the wort into the kettle.

4. Bring the wort to a boil. Boil it for 45 minutes. Add the hops and the Whirlfloc tablet and continue to boil for 30 minutes. Put your wort chiller into the wort at least 15 minutes before the end of the boil.

5. At the end of the boil, add the knockout hop addition, but don't start chilling yet. Cover with a lid or clean trash bag and let the wort sit hot for 30 minutes, then chill to 65°F.

6. Siphon the wort into your sanitized fermentor and pitch two packs of liquid yeast or a 2-liter starter.

7. Ferment at 65°F for two weeks, adding the wood chips after primary fermentation slows down (usually after five to seven days).

8. Keg or bottle the beer. (If you're bottling, I recommend 3¾ ounces of dextrose/corn sugar for this beer.)

BREWING WILD AT HOME

When making wild beer at home, you need to have an open mind. You give control of fermentation to the microbes you choose and the ones that choose you. Whether you buy a commercial culture, culture one of your own, or pitch bottle dregs, chances are that the resulting beer will taste unlike anything you can buy at the store. When it turns out to be fantastic, your friends will be impressed that you brewed one of the most difficult beers in all of brewing.

Remember that any plastic equipment (siphon tubing, airlocks, buckets, and so on) that comes in contact with wild beer should be labeled as such and used only for wild beers thereafter. Also, make sure you have plenty of patience!

CAPTURING YOUR OWN CULTURE

If you really want to get in touch with your inner localvore, try a spontaneous fermentation with your own town's bugs. You're getting into uncharted territory here. More than likely, you're going to make some really funky stuff. But you're also going to be one of the elite group of homebrewers who is blazing a trail.

YOU NEED

1½ gallons wort at 1.030 (can be made from pale dry extract)

3 (2-liter) soda bottles, emptied and cleaned

sanitizer

funnel

cheesecloth

rubber bands

The Basics

You can capture wild yeast and other microbes almost anywhere in the world. There's no way to get consistent, repeatable results from just the air, but you can try to stack the deck in your favor.

First, time it right. Most lambic breweries take the summer off. American brewers with coolships, such as Allagash, tend to do the same. Whether it's due to a higher percentage of undesirable microbes or a longer danger zone while the wort cools slowly in the higher ambient temperature, avoid brewing wild in summer. Throughout much of the world winter also isn't the best time to capture wild microbes. Many brewers say it's due to the opposite effect of summer: Temperatures below or near freezing aren't ideal for the microbes we want in beer. Stick to fall and spring for these experiments.

You can control the season in which you capture microbes, but you can't control what you capture. Make up for this by increasing your experiment size. Try at least three locations around your house to see what happens.

Start Small

Start by sanitizing your three 2-liter bottles. Then pour the 1.030 wort through a funnel to distribute it equally among the bottles, leaving 4 to 5 inches of headspace at the top of each. Wrap the cheesecloth around the top and use a rubber band to keep the cheesecloth in place.

Place each bottle in a different location in your house or on your porch. I recommend areas with good circulation but different atmospheres. You might leave one in the basement, one upstairs by a window, and one on the porch (where animals can't get it). If you live in an apartment, try one at your place and recruit a couple of friends who live nearby and leave bottles at their places. Again, access to open air is usually good, so leaving a bottle by a window is ideal. Leave the bottles out overnight.

The next morning, loosely cap them—but not to the point that they seal—and place them in a cool (60–70°F), dark spot. After a few days, you may see some activity, but leave them for two to three weeks to see what happens. At that point, you can give each one a smell and see what's going on. Do not taste them! As fermentation continues and the wort ages, anything really nasty will die off. But at this point, you just have something growing in rich, sugary water. Drinking it could prove dangerous.

If you smell rancid or garbage-like aromas, the beer can go right down the drain. If you smell fruity or spicy aromas, it could be a keeper and worth developing. You have two options: Tighten the cap, checking periodically on the pressure and off-gassing it yourself, or attach an airlock.

In three to six months, the beer should have a more distinctive aroma one way or the other. If you take a gravity sample and have reached a significantly lower gravity (say, 1.015 or below), go ahead and taste it as well. If you like what you smell and taste, you now have a house culture to pitch in your next beer! Pour off most of the clear wort and make a starter with the dregs (page 110).

Variation: Take a Bigger Gamble

If you want to roll like the Belgian lambic brewers, you need to coolship your beer. Coolshipping entails cooling your beer, open to the chilly night air, overnight. Take your wort after boiling and set it outside, covered with some cheesecloth to keep out bugs and leaves. Time of year is important: Late fall to early spring is best. Summer is bad, and, depending on your location, the middle of winter is probably too cold. By the next morning, your wort should be chilled to 50–70°F. At that point, rack into a carboy, attach an airlock, and forget about it for a year. The only thing you can do wrong is letting the airlock dry out during aging. Other than that, it's up to the wild bugs in the air to do their jobs. (See page 130 for more info.)

This technique, while completely authentic, is unfortunately fraught with failure. I usually have to dump 4 of 5 batches before I get one that tastes great. Once I have a good one, though, I can repitch that yeast and have my own native wild culture unlike what any other brewer has. The majority of barrels at my brewery contain 100 percent native wild cultures with no cultured yeast added.

Variation: Try It with Fruit

You know that white film on the outside of wild grapes and figs? That's yeast! Putting a handful of grapes into a small jar of wort usually will start fermentation pretty quickly. Again, if the starter smells good, then you can grow it into a bigger batch. From a tree in my front yard, I recently picked six plums covered in yeast and put them into a jar of wort. Within a few hours, it was fermenting away. Imagine my surprise when, sampling the wort after a week, it tasted exactly like a German hefeweizen! It had mild clove notes in its aroma and taste, with just a hint of tartness. It was otherwise very clean. Now I have a tasty yeast to use that nobody else can order from a lab. The lesson: It may not be sour, but it will be unique.

CULTURING MICROBES FROM SOURS

One of the easiest ways to kick-start your sours program is to get some bugs from the best. That's right: If you like a commercial sour beer, there's a good chance you can use their cultures in your own beer.

Jolly Pumpkin is a great source for fresh bugs, but I've also gotten healthy starters going from several of the Belgian lambics. In general, the more bottle dregs you add, the better. Also, recently packaged beer will have more active cultures—cultures that are alive and at work in the beer itself. Lower alcohol content is usually good, too. You do need to make sure there are active bugs, though. A few breweries that make tasty sours, such as New Belgium, pasteurize their beers, which means there's nothing alive in the bottle.

YOU NEED
several bottles sour beer

small jar (2–3 ounces)

medium jar (6–12 ounces)

growler or other 2-liter container

aluminum foil (if the jars don't have lids)

sanitizer

1 gallon filtered brewing water (page 12)

2/3 pound light dry malt extract

1/4 teaspoon yeast nutrient

large pot

large glass

A FEW RECOMMENDED BREWERIES FOR DREGS

The Bruery

Jolly Pumpkin

Lost Abbey

Russian River

Almost any Belgian brewer of fresh bottled lambic or gueuze (except Lindemans)

Note: The amount of yeast in a single bottle of beer is very small, and you shouldn't stress it by starting it off in a large starter. We'll start it off in a small jar with about ¼ cup of wort. After a week, step it up to 2 cups. Finally, after another week, step it up to 2 liters.

1. A few days ahead of time, make sure the bottles of beer from which you're going to culture the yeast are stored upright. Anything that drops from suspension will be located conveniently at the bottom of the bottle.

2. In the meantime, sanitize your jars by soaking them in sanitizing solution or by boiling them for 15 minutes.

3. Boil the dechlorinated water, dry malt extract, and yeast nutrient in a pot for 30 minutes. Pour the boiling wort into the jars, leaving a decent headspace, screw on lids or wrap foil over the tops, and allow the jars to cool.

4. Crack the bottle open and pour the beer into a large glass, reserving the last ½ inch or so of beer. Leave more or less if you can see the sediment, but err on the side of more. Pour slowly to minimize sloshing. Now swirl the yeast into solution and sanitize the lip of the bottle.

5. Crack the lid on the smallest jar of starter wort, quickly pour the sediment into it, and replace the lid. If you have multiple bottles, repeat the process.

6. You won't see much activity in the first jar, perhaps just a thicker layer of sediment on the bottom than what you started with. After one week, swirl any sediment into solution and pour it into the next size jar of starter wort. This time, you should see some activity, such as some krauesen (foam) on top and a growing layer of sediment on the bottom.

7. After another week, step up the starter to the largest jar. This should be ready to pitch into your 5-gallon batch within a few days, but I prefer to let it ferment to completion over another week and then sample the wort before I brew up a 5-gallon batch. It should smell and taste similar to the beers that you harvested the yeast from, but more subdued.

USING BRETT

Although some brewers think of Brett as a bacteria or a contaminant, it's merely another yeast. It originally was isolated from British ales and probably was present in a large amount of beer until the 1900s. It's a quirky yeast compared to *Saccharomyces*. It ferments fairly cleanly when pitched like a primary yeast, but it produces the flavors we want only when treated badly.

When fermenting with Brett under ideal conditions, it's very neutral and clean—it would be hard to know that the beer was fermented with anything other than a traditional ale yeast. There's no tang or funk, just a slightly estery beer similar to an aggressive British ale strain such as Yorkshire.

If you want Brett character, the best time to add it is when the beer has few to no simple sugars left, the pH is low, and alcohol is present. Then Brett will thrive and produce the flavors we appreciate. It's also best under pressure (literally) and frequently will produce the flavors you want when added at bottling. So, while it's interesting to make a big starter using Brett as your primary yeast, it's better to use it at bottling and to give it several months to develop complexity.

KETTLE SOURING

In the previous edition of this book, I recommended sour mashing to make a quick sour. I wasn't a big fan of the technique, but it was the best way at the time. Since then, a revolution has taken place in the way that homebrewers and commercial brewers make quick sour beers—called "kettle souring."

Kettle souring is the process of taking boiling wort with no added hops and chilling it to around 90–100°F. Then you add some Lacto, normally in the form of probiotic pills (see below), probiotic drinks such as Goodbelly, or a commercially produced pitch. After 24–48 hours, the wort is nicely sour, and you bring it to a boil to kill the Lacto. This method prevents the Lacto from infecting your equipment and locks in the level of sourness. At this point, you can add hops if you want, chill it to 65°F, pitch a healthy ale yeast, and ferment as normal. When the fermentation finishes, you can add fruit or dry hops if desired or just keg and bottle. You will have a two-week-turnaround sour beer that will impress your friends.

This technique has a much higher success rate than the old sour-mash technique, and you have a few ways to ensure success. Oxygen is the enemy of sour beer, and once your wort cools you should minimize any oxygen intake while the Lacto is working. The best way to do this is to purge a corny keg with CO_2 (page 20), then siphon your wort into it and add the Lacto. If you don't keg, placing a layer of plastic wrap over the top of the wort will keep oxygen out. The secret pro technique is to lower the pH of the wort to 4.5 with lactic or phosphoric acid before adding the Lacto. This requires a pH meter and isn't essential, but it does help with foam stability and keeping undesirable bugs from growing.

Sources of Lactobaccilus

With the popularity of probiotics, there's a plethora of sources for Lacto as close as your local grocery store. Look in the yogurt section of your local supermarket for Goodbelly probiotic drink. Loaded with *Lactobaccilus plantarum*, it comes in different flavors and also in "shots." One quart of the flavored drink is plenty for a 5-gallon batch. (The added fruit flavors don't come through in the final beer, but, if you're worried about that, find the unflavored shots and use 4–5 in a 5-gallon batch.)

Probiotic pills are another cheap and easy source of Lacto and usually are available at your local supermarket. If you need an online source, check out swansonvitamins.com and search for "plantarum." A jar of 30 capsules costs less than $10 and is plenty for a 5-gallon batch. Overpitching isn't a problem when kettle souring, so I usually open all 30 pills and toss the powder into the fermenter, but you could use 10 pills and be fine.

Yogurt is another easily available source of Lacto. Look for a nonfat, unpasteurized variety. Most fat-free Greek yogurts will work fine. Add around ½ cup of yogurt directly to the fermenter.

Most yeast labs sell Lacto, but for the price you're better off with one of the above choices.

Lactobacillus plantarum is the most popular strain for kettle souring because it works best at 80–90°F, whereas other strains (such as in yogurt) work best at 120°F, which can prove difficult for homebrewers to maintain for 48 hours. *L. plantarum* also is extremely hop sensitive. As little as 1 IBU can keep it from working, so don't add *any* hops until you're done souring.

THE FISH AND THE RING BELGIAN ALE (WITH BRETTANOMYCES)

The brewers at Orval add a mixed culture containing mostly *Brettanomyces* yeast at bottling. The beer is dry hopped with Styrian Goldings so that it can have a bright hop aroma with a little Brett character when it's fresh from the brewery. Over the course of a year, the hop aroma fades and the perfumey wild character of the Brett starts to prevail. This beer isn't that difficult to brew compared to many "wild" styles, and tasting the evolution of this beer over the course of a year is enlightening.

YOU NEED
basic brewing equipment (page 3)

9 gallons filtered brewing water (page 12)

9 pounds German pilsner malt (81.8%)

1¼ pounds CaraMunich II (11.4%)

¾ pound dextrose or corn sugar (6.8%)

10 alpha acid units Hallertau hops at 60 minutes (39.6 IBU)

1 Whirlfloc tablet

3¾ ounces dextrose/corn sugar

2 vials or packages Bastogne Belgian Ale WLP510 yeast (or a 2-liter starter made from one pack; page 110)

1½ ounces Styrian Goldings hops (dry hops)

1 vial or pack Brettanomyces bruxellensis WLP650/WY5112 or the dregs from 3 bottles Orval (freshest you can find)

TARGETS
Yield: 5 gallons

OG: 1.058–1.061

FG: 1.010–1.012 (will go lower in the bottle)

IBU: 39.6

1. Mix the malt with 4 gallons of water at 165°F or the appropriate temperature to mash at 150°F. Mash for 60 minutes.

2. Recirculate the wort until it's fairly clear. Run off the wort into the kettle.

3. Sparge with 5 more gallons of water at 165°F. Run off the wort into the kettle.

4. Bring the wort to a boil. Boil it for 30 minutes. Add the Hallertau hops and continue to boil for 60 minutes. Add the Whirlfloc tablet and the dextrose or corn sugar at 30 minutes. Put your wort chiller into the wort at least 15 minutes before the end of the boil.

5. When the boil finishes, cover the pot with a lid or a new trash bag and chill to 68°F. Siphon the wort into your sanitized fermentor and pitch two packs of liquid yeast or a 2-liter starter.

6. Ferment at 65°F for two days, then let the temperature rise on its own to 72–75°F. Add the dry hops after primary fermentation slows down (within six to seven days).

7. Bottle the beer with 3¾ ounces of dextrose/ corn sugar. Add the *Brettanomyces* culture (or the dregs of the bottles of Orval) at the same time as the priming sugar. After bottle conditioning, leave the bottles at cellar temperatures (50–60°F) so that the wild yeasts can continue to produce interesting aromas and tastes over the coming year.

BERLINER WEISSE

This is the basic process for making a quick kettle-soured beer, and variations follow at the end for making fruited sours or goses.

YOU NEED

basic brewing equipment (page 3)

8 gallons filtered brewing water (page 12)

4½ pounds pilsner malt

3 pounds wheat malt

1¾ alpha acid units Hallertau hops at 30 minutes (4 IBUs)

1 quart Goodbelly probiotic drink, 4–6 shots, 10–30 probiotic pills, or ½ cup fat-free, unpasteurized yogurt

2-liter starter California Ale WLP001/American Ale WY1056 or 2 packs US-05 dry yeast rehydrated

7 ounces dextrose/corn sugar (optional, use only for bottling)

TARGETS

Yield: 5 gallons

OG: 1.035–1.037

FG: 1.005–1.007

IBU: 4

1. Mix the malt with 3 gallons of water at 165°F or the appropriate temperature to mash at 150°F. Mash for 60 minutes.

2. Recirculate the wort until it's fairly clear. Run off the wort into the kettle.

3. Sparge with 5 more gallons of water at 165°F. Run off the wort into the kettle.

4. Bring the wort to a boil, then turn the heat off. Add your wort chiller and let it sit in the hot wort for 15 minutes. Cover the wort with a lid or clean trash bag and chill to 90°F if using *Lacto plantarum* or to 120°F if using yogurt.

5. Rack into your fermenter, preferably a CO_2-purged keg or at least gently siphoned into a carboy or bucket, and add your Lacto. If fermenting in a bucket, place a layer of plastic wrap over the top of the wort to prevent oxygen uptake.

6. After 24 hours, take a small sample and taste it, keeping in mind that the sugar in the wort will offset some of the acidity. If it's in the range of sourness that you like, proceed to the next step. If not, give it another 24 hours.

7. Bring the soured beer to a boil, add the hops, and boil for 30 minutes. Put your wort chiller in the wort at least 15 minutes before the end of the boil.

8. When the boil finishes, cover the pot with a lid or a new trash bag and chill to 65°F. Siphon the wort into your sanitized fermenter and pitch the rehydrated dry yeast or the 2-liter starter.

9. Ferment at 65–70°F for 2 weeks.

10. Keg or bottle the beer. (If bottling, use 7 ounces of dextrose/corn sugar for this beer.)

VARIATIONS

Gose: Add 20 grams of freshly ground coriander and 20 grams of salt to the kettle at the end of the boil and proceed as indicated.

Fruited sours: After fermentation completes, add 5–10 pounds of the fruit of your choice (page 177), and wait 3 weeks before packaging.

DIY LAMBIC

Brewing a traditional lambic requires cooling wort, while open to the chilly night air, overnight (aka coolshipping), then aging it for at least a year while the wild yeasts and bacteria slowly work their magic. Nothing is stopping homebrewers from attempting this spontaneous style of fermentation, but it does have a high rate of failure. Most of my attempts have a strong aroma and taste of rubber bands, but a few batches have come out fantastic. In the next chapter, we'll discuss the best ways to make lambic-style beers, both with the traditional method as well as simpler and more successful methods. If you lack the patience or space to let these beers age for a year or more, try a quick sour such as the Berliner Weisse (page 147) or American Wild Ale (page 152).

A FEW NOTES:

➡ The traditional lambic recipe consists of 35 percent raw wheat and 65 percent pilsner malt. Using raw wheat requires cereal mashing (page 161), so unless you're comfortable doing a cereal mash, just substitute malted wheat or flaked wheat. The flavor difference in a sour beer will be minimal. Remember to add an extra pound or two of malt to hit your OG when using raw or flaked products (page 160).

➡ Aim for a gravity of 1.050 with just a pinch of hops (around 5 IBU). I don't use smelly old whole hops like the traditionalists do because it can take hours of boiling to drive off the stinky-foot aroma. As long as you use a mild, neutral, low-alpha hop (such as Willamette), you'll be fine. Whatever you do, don't increase the IBU above 10 because that can inhibit some wild cultures.

➡ Mash higher than usual, shooting for as high as 159°F. Brett and other cultures will bring the gravity down lower than typical brewing yeast, and the bugs will feed off the long-chain sugars produced by a high-temperature mash. Try to get a lot of cloudy runoff from your mash into the boil, too, because the bugs will feed on these starches while the beer ages.

➡ Invest in duplicates of all equipment that touches the beer after pitching (bucket, racking equipment, bottling equipment, and tubing) unless you're willing to dedicate your current equipment to sours only.

YOU NEED

basic brewing equipment (page 3)

8½ gallons filtered brewing water (page 12)

3½ pounds pilsner malt (31.8%)

3½ pounds North American 2-row malt (for extra enzymes and husks) (31.8%)

4 pounds raw wheat or flaked wheat (if using raw wheat, page 160) (36.4%)

1.25 alpha acid units Willamette hops at 60 minutes (4 IBU)

If you're attempting a spontaneous fermentation, you won't need any yeast. Otherwise, you'll need . . .

bottle dregs from a couple of your favorite sour beers grown into a healthy 2-liter starter (page 110). Alternatively, you can make a starter from a commercial culture that includes a blend of *Saccharomyces*, *Brettanomyces*, *Lactobacillus*, and *Pediococcus*, such as Sour Mix 1 WLP655 or Belgian Lambic Blend WY3278.

7 ounces dextrose/corn sugar (optional, use only for bottling)

TARGETS
Yield: 5 gallons
OG: 1.050–1.053
FG: 1.003–1.007
IBU: 4

1. Mix the malt with 3½ gallons of water at 175°F or the appropriate temperature to mash at 158–160°F. Mash for 60 minutes.

2. Sparge with 5 more gallons of water at 190°F. Run off the wort into the kettle. Don't recirculate until clear before running off.

3. Bring the wort to a boil, boil for 30 minutes, add the hops, and continue to boil for 60 minutes. Put your wort chiller into the wort at least 15 minutes before the end of the boil.

4. When the boil finishes, you have two choices. If you want a traditional spontaneous fermentation, put the hot wort outside, cover it with a layer of cheesecloth, and let it cool overnight. The next morning, siphon it into a sanitized fermenter. If you're pitching yeast, chill the wort to 65°F, siphon it into your sanitized fermenter, and pitch the starter of mixed cultures.

5. The choice of fermenters is up to you. Plastic buckets are nice because they're disposable, but, when it comes to oxygen intake over the course of a year, they suck. The more oxygen you take in over the year or two of aging, the more it can cause your beer to turn slowly to vinegar. I recommend a glass carboy or one of the new plastic carboys. Aging in a keg seems like a good idea, but you want a small amount of oxygen to enter the beer, so stainless isn't the best choice for wild ales. The worst thing you can do is let the airlock dry out. Check it monthly at least.

6. After six months, taste it. You may see a layer of what looks like white mold on the top of the beer. Don't fret— it's only the pellicle, a layer of wild yeasts that keeps oxygen out. (See photo below.) Remove a small sample and give it a smell and a taste. The aroma should be pleasantly funky and may or may not taste sour. Sourness often develops last and can take as long as 12–16 months. If you're fermenting a spontaneous beer, this can be a make-or-break moment. If it smells horrible or like plastic, dump it and try again.

MIXED-CULTURE FERMENTATION

Using a mixed culture is even easier than kettle souring because you don't reboil the wort after souring. You simply pitch all your bugs into the fermenter and let them fight it out. The only problem with this method, unlike kettle souring, is that it can infect all your equipment, but this is more of a worry for commercial brewers. As long as you have separate hoses and any other plastic items, you should be fine.

This technique starts with a quick fermentation with ale yeast and then adding Lacto and/or Brett. This method can create a nice and fairly complex sour beer in as few as 3–4 weeks. Just remember that Brett will eat complex sugars slowly over time, so make sure the OG is below 1.005 if bottling.

Another cool trick is to halt the Lacto when it gets to your desired level of sourness by dry hopping. An ounce or two of dry hops will stop *Lactobacillus plantarum* dead in its tracks. I highly recommend any New Zealand variety for this style of beer, especially Nelson Sauvin.

BASIC TECHNIQUE

Brew a simple wheat beer with no hops, OG 1.050, 70 percent pilsner and 30 percent wheat.

Chill to 65–70°F and siphon into your sanitized fermenter

Pitch the *Saccharomyces* strain of your choice.

After 3 days, pitch *Lactobacillus* in the form of 1 quart of Goodbelly probiotic drink, 4–6 Goodbelly shots, or 10–20 opened probiotic pills. Also add *Brettanomyces* at this time if desired.

Add dry hops, if desired, when acidity reaches the level you want.

Wait until the FG falls at least below 1.005 if using *Brettanomyces* or below 1.010 if using just Lacto.

Package as desired, aiming for 3.2 volumes if bottling (6.3 ounces of dextrose or corn sugar).

Note: If you keep the fermentor warm (70–85°F), you should have a nice funky beer with decent sourness within 6 months. You can accentuate the sourness with a teaspoon or two of lactic acid or acid blend at bottling, too. It probably won't be quite up to the quality of a sour that had more time to develop, but you can weigh the risk and the reward and decide what's right for you.

7. Check it again after a year. If the beer isn't sour enough for your liking, let it sit for another 3–4 months and sample it again, or adjust the sourness with food-grade lactic acid or a winemaker's acid blend—a teaspoon or two should do the trick. If it's nice and tart, you're ready to bottle. (This really isn't a beer you want to keg.)

8. Rehydrate some dry white wine or Champagne yeast, add it to your beer along with your priming sugar solution, and bottle the beer. I recommend using 7 ounces of dextrose/corn sugar for this beer. It's traditional to carbonate this type of beer at even higher volumes, but increase the priming sugar at your own risk—and using only Champagne-style bottles. Brett easily can consume too much sugar, especially if the beer wasn't totally finished fermenting. (It should be 1.007 or lower.) Bottle bombs are a real danger.

9. Even after bottling, the bugs may continue to work very slowly over the years, so sample a bottle every month or two to make sure the bottles aren't over-carbonating. If they are, store them in the fridge or drink them.

AMERICAN WILD ALE

This wild beer has a healthy dose of American craft-brewing attitude.
The funky Brett blends well with the complex caramel and hop tastes.
It doesn't take as long to brew as a lambic, since strong sourness isn't the goal.
You'll be able to drink it within two to three months, but it'll continue
to evolve for several years if you let it sit.

YOU NEED

basic brewing equipment (page 3)

10 gallons filtered water (page 12)

15 pounds North American 2-row malt (96.8%)

½ pound Weyermann CaraAroma (3.2%)

5 alpha acid units Amarillo hops at 60 minutes
(17 IBU)

1 ounce Amarillo hops (dry hop)

1 Whirlfloc tablet

1 vial or package California Ale WLP001/
American Ale WY1056 yeast and a small
starter (500 milliliters) the dregs from a
bottle of sour beer (page 110)

1 ounce Citra hops (dry hop)

2 ounces medium-toast French oak cubes or
chips, steamed for 10 minutes, then added to
the fermentor after 1 week

5 ounces dextrose/corn sugar
(optional, use only for bottling)

TARGETS
Yield: 5 gallons
OG: 1.066–1.068
FG: 1.010
IBU: 17

1. Mix the malt with 5 gallons of water at 165°F or the appropriate temperature to mash at 150°F. Mash for 60 minutes.

2. Recirculate the wort until it's fairly clear. Run off the wort into the kettle.

3. Sparge with 5 more gallons of water at 165°F. Run off the wort into the kettle.

4. Bring the wort to a boil. Boil it for 15 minutes. Add the Amarillo hops and continue to boil for 60 minutes. Add the Whirlfloc tablet at 30 minutes. Put your wort chiller into the wort at least 15 minutes before the end of the boil.

5. When the boil finishes, cover the pot with a lid or a new trash bag and chill to 70°F. Siphon the wort into your sanitized fermentor and pitch the yeast and the starter of the sour beer dregs.

6. Ferment at 70–75°F for three weeks, adding the dry hops and oak chips or cubes after primary fermentation slows down (usually within five to seven days).

7. Keg or bottle the beer. (If you're bottling, I recommend 4 ounces of dextrose/corn sugar for this beer.) After a month in the bottle or keg, the resulting beer should be a light to medium funky, with dank hops, a complex caramel taste, and just a hint of warm, toasty oak.

OTHER FERMENTABLES
and
TASTES

UNMALTED STARCHES

Unusual or unmalted grains such as quinoa, buckwheat, and spelt make up a fermentable category that has become more and more popular with brewers, particularly for specialty beers. This shift away from Reinheitsgebot (see below) has sped along with the extra enzymes that abound in today's North American base malts and in many pilsner malts from Europe. Their ability to convert the starch in non-barley ingredients means you generally can add up to 50 percent starchy adjuncts to the mash and have them converted into fermentable sugar. Quinoa beer with hefeweizen yeast, a stout brewed with 50 percent oats, and tasty buckwheat red ale are all possibilities.

Almost all of the ingredients discussed in this section should be crushed finely before mashing and heated in water prior to the rest of your mash if the gelatinization temperature is higher than the mash temperature. (See "Cereal Mashing" on page 161 before you start.) If you don't want to add 30 minutes to your brew day, many of these starches come in flaked form. (Flaked grains are raw grains that have been wet down and passed through hot rollers, which gelatinize the starch.) You also can use any grain in flour form.

On the following pages you'll find a list of some common starchy adjuncts, but by all means try something new if it has starches to convert! I've succumbed to the excitement that comes from making a beer from an unusual starch. I've made water-chestnut beer, tapioca beer, the inevitable pumpkin beer, even red-beans-and-rice beer. (I don't recommend that last one.) The flavor contribution from almost all of these fermentables isn't as intense as you would think. Often, you'll get only a subtle nutty or grainy flavor. A good rule of thumb is to start with at least 20 percent of your ingredient in the grain bill to get a notable flavor contribution, but cap it at 40 percent in case it gums up the mash. Also, just because your beer doesn't showcase one particular ingredient doesn't mean you should forget about it. Many of these nontraditional fermentables will add complexity to your beer, and they're certainly worth trying.

REINHEITSGEBOT

Some respect it, some scoff at it, but the Bavarian brewing law known as the Reinheitsgebot is sure to start an argument among brewers. It originated in 1516 and stated that only barley, hops, and water were allowed in beer. (Yeast was still a mystery at that point.) The law was enacted primarily to stop brewers from adding dangerous herbs and additives prevalent at the time, but it also helped to reserve more rye and wheat for bakers. Initially, the law affected only Bavarian brewers, but it eventually spread throughout Germany. Later on, it became legal for German ales (but not lagers) to contain wheat and other grains as well as sugars in dark versions.

For modern brewers, following Reinheitsgebot requires some sneaking around, because no acid is allowed for adjusting the pH of the water. No mineral additions, finings, or enzymes are allowed, either. Some brewers have the pH of the water adjusted before it enters the brewery, whereas others use a malt that has lactic acid on the grain to lower the mash pH. (See "Acid Malt," page 37.) Keeping harmful ingredients from beer helped create a quality product, but it also minimized creativity in Germany. The decision still affects the brewing culture there: Most breweries make variations of the same four or five styles, with some regional differences.

GREAT GRAINS

Rice: This grain is a traditional adjunct used in North American and Asian lagers. Together with corn, it can make up to 50 percent of a grain bill for some big breweries! But rice has little flavor, so why do large breweries use so much of it? Megabreweries use 6-row malt, which has high protein levels, so the rice (and corn) dilutes the high protein levels, which, if left alone, will create clarity problems in the finished beer. Rice also lightens the flavor and body of beer. For craft brewers, this can be useful in Belgian tripels and double IPAs, where you want a crisp, clean taste. If you want to give that a try, use rice by dry weight for 10–20 percent of your grain bill, normally 1½ pounds for a 5-gallon batch.

Rice needs to be gelatinized before mashing, so cook it as if you were going to eat it before adding it to the mash. Some brewers have had luck adding instant rice directly into the mash. It's precooked, so it should be fine. Rice flakes also can be added directly to the mash, but some brewers claim they produce an oxidized flavor.

Don't overlook the wide variety of rices in the marketplace: perfumey jasmine rice, nutty brown rice, black "forbidden" rice, and American wild rice, among other options. Just cook them according to the directions and add them to your mash.

Corn: Another traditional adjunct used in some North American beers, corn does contribute flavor, but it's subtle and notable only when used as more than 20 percent of the grain bill. Using 15–30 percent is common for beers such as cream ales or American light lagers. You can buy flaked corn at most homebrew supply stores. Flaked corn is pre-gelatinized, so you can add it right to the mash.

If you want to use corn grits, currently the cheaper option, cook them as if you were going to eat them before adding them to the mash (but don't add any butter). Brewing with fresh corn cut from the cob is surprisingly disappointing, because the corn flavor is hardly detectable. Don't confuse the flavor of corn with dimethyl sulfide (DMS), a cooked corn aroma and taste derived from malt. Rolling Rock takes pride in the high level of DMS in its beer, but most brewers consider it a fault. A vigorous boil for 75 minutes usually cooks all of the DMS from the wort.

Oats: You can buy malted or unmalted oats, but since traditionally they're used in only a few styles, they belong here with other adjuncts more than with malted barley. Oats are supposed to give beer a silky mouthfeel, but that depends on the oats. I brewed an oatmeal stout with 80 percent malted oats, and they didn't improve the beer's body. That said, oats, like rye, contain a large amount of beta-glucans, which are viscous, long-chain, unfermentable sugars that add a slick, oily mouthfeel to a beer. This can help add perceived body and is likely why oatmeal stouts, traditionally fairly low in alcohol, have some oats in the mash. As with flaked corn, you can use flaked oats if you don't want to cook the oats before you use them. Oats have little or no flavor, so toasting them in the oven at 350°F for 15 minutes can bring out some subtleties. Use at least 10 percent toasted flaked oats if you want any chance of noticing oats in the finished beer.

Raw Wheat: Raw wheat has a notably tangy flavor compared to malted wheat and is essential for brewing authentic witbiers. Commercial brewers rarely use it except in Belgian witbiers or lambics. Most of the beer literature will tell you that raw wheat gelatinizes at normal mash temperatures, but I haven't found this to be the case. If you're worried about it, do a cereal mash as you would for the other adjuncts or use wheat flakes instead. Belgian witbiers commonly use 40–50 percent raw wheat, while traditional lambics use about 30 percent.

Buckwheat: Although they sound similar, buckwheat and wheat aren't related. Buckwheat is a seed from the rhubarb family. It doesn't have much taste, but toasting it will bring out some of its flavor. Grocery stores sell pretoasted buckwheat as kasha—which you usually can find it in the international section by the matzo crackers. Cook it according to the package directions and add it to the mash. As with most specialty starches, use at least 20 percent in the grain bill or you probably won't taste it. Don't go higher than 40 percent because you could end up with a stuck mash. Use North American 2-row or 6-row malt to ensure that the mash has enough enzymes.

Raw Barley: Grocery and health-food stores sell unmalted barley. It can be boiled and added to the mash as with many other starches. Since it's barley, it won't bring anything interesting to the table. It will add beta-glucans, however, which can help the mouthfeel of a beer. Raw barley in flaked form often goes in Irish stouts in the 5–10 percent range.

Spelt: Spelt is an ancient variety of wheat and has a similar flavor. The Weyermann malting company makes a malted spelt, which you can use for up to 60 percent of a grain bill. Raw spelt is a popular ingredient in saisons and other farmhouse ales, though it's typically 30 percent or less of the grain bill. I don't think that spelt tastes much different from wheat.

Quinoa: This highly nutritious grain has a pleasant nutty flavor. When used in a delicate style, such as a Belgian witbier, in the 30 percent range, it can impart a nuttiness beyond that of grains such as raw wheat. Quinoa has a high protein level (14–18 percent), though, which will cause some haze issues but should help head retention. Like rice, it should be cooked fully before you add it to the mash.

PROTEIN LEVELS

The lower the protein levels in brewing grains, the better. Protein can help with foam stability, but high levels of protein can cause haze problems, reduce mash efficiency, and result in stability issues. High levels of protein also imply high levels of beta-glucans, which can cause a sticky mash. Brewers can degrade some of the beta-glucans and protein by performing a beta-glucan rest in the 115–117°F range for about 30 minutes.

North American 2-row malt tends to have around 11–12% protein and benefits from a shorter (15-minute) protein rest around 122°F. North American 6-row malt is higher in protein, at around 13%, and a 30-minute rest at 122°F is recommended for clarity. European barley is a little lower in protein, but British barley is the lowest, at around 10%. Protein rests for British malts are unnecessary and may degrade the protein to a level where you start losing head retention.

Some of the alternate grains and starches in this section can have high protein levels, which should inform your choice of mashing regimen and base malt selection. But you need high levels of enzymes to convert the extra starch, so low-protein malts such as British pale ale aren't appropriate. Stick with a North American 2-row malt with the lowest protein level you can find.

RAW WHEAT EXPERIMENT

Over the years, I've brewed dozens of witbiers. Whenever I used raw wheat at the traditional 40–50 percent the grain bill, I always had really low starting gravities. The problem went away for the most part when I boiled the raw wheat for 30 minutes before adding it to the mash. According to common wisdom, though, this shouldn't be necessary. Since wheat gelatinizes between 125°F and 145°F, you should be able to add it to a typical mash and have it gelatinize. I decided to mash raw wheat in various forms and measure the results.

Control Batch with Malted Wheat

This batch set the basic efficiency of my system.
5$\frac{1}{2}$ pounds pilsner malt
4$\frac{1}{2}$ pounds malted wheat
Mash: 154°F (75 minutes)
OG: 1.041 (6$\frac{1}{2}$ gallons); 71 percent efficiency

Standard Infusion with Raw Wheat

This should be the same as the gravity above, but obviously it isn't.
5$\frac{1}{2}$ pounds pilsner malt
4$\frac{1}{2}$ pounds raw wheat
Mash: 154°F (75 minutes)
OG: 1.026 (6$\frac{1}{2}$ gallons); 45 percent efficiency

Upward Infusion (Protein Rest) with Raw Wheat

A protein rest helps make the starches more available, but we're still not back to the control batch's efficiency.
5$\frac{1}{2}$ pounds pilsner malt
4$\frac{1}{2}$ pounds raw wheat
Mash: 118°F (20 minutes), 154°F (60 minutes)
OG: 1.030 (6$\frac{1}{2}$ gallons); 52 percent efficiency

Standard Infusion with Flaked Wheat

Flaked wheat supposedly is pre-gelatinized by the flaking process, yet efficiency still took a substantial hit.

5$\frac{1}{2}$ pounds pilsner malt
4$\frac{1}{2}$ pounds flaked wheat
Mash: 154°F (75 minutes)
OG: 1.032 (6$\frac{1}{2}$ gallons); 55 percent efficiency

Cereal Mash with Raw Wheat

This method yielded results substantially closer to the control, but you still have to add additional grain to hit your target gravity using this method.
5$\frac{1}{2}$ pounds pilsner malt
4$\frac{1}{2}$ pounds raw wheat
Mash: Heat $\frac{1}{2}$ pound of pilsner malt and the raw wheat to 154°F. Hold for 15 minutes, bring to a boil, and simmer for 20 minutes. Add the rest of the pilsner malt and water so that the temperature stabilizes at 154°F. Hold for 75 minutes.
OG: 1.036 (6$\frac{1}{2}$ gallons); 62 percent efficiency

Standard Infusion with Wheat Flour

While it made for a slow run, this method resulted in the highest extraction any method, even outperforming the control batch with malted wheat.
5$\frac{1}{2}$ pounds pilsner malt
4$\frac{1}{2}$ pounds wheat flour
Mash: 154°F (75 minutes)
OG: 1.044 (6$\frac{1}{2}$ gallons); 76 percent efficiency

Conclusions: You can't count on raw wheat or other raw grains to give you normal points per pound per gallon (PPG). The different mashing methods varied, but raw flour yielded the highest extraction, even without a cereal mash or protein rest, which suggests that milling is the most important factor. When using raw wheat or other grains, grind as finely as your mill will allow—it may take multiple passes—and add an extra pound or so of base malt to ensure you hit your starting gravity.

CEREAL MASHING

Using starches other than malted barley isn't difficult, and it may not even add time to your brew day. Just make sure that your starch reaches its gelatinization temperature so that the starch is available for conversion into sugar by the enzymes in the mash. For some starches, that may mean simply stirring it into a typical mash (145–160°F), whereas for others it will mean boiling for up to an hour (see below). The milling of the starch is very important. Big chunks of grain won't gelatinize as easily as starches ground into a flour. There's no harm in boiling a starch to ensure complete gelatinization, so when in doubt go for a higher temperature.

Use at least 60 percent North American 2-row or 6-row malt in your recipe's grain bill to convert the nonmalted starch or starches.

YOU NEED

basic brewing equipment (page 3)

3- to 5-gallon pot (for some starches only)

filtered brewing water (page 12) in a ratio appropriate for your starch (see chart below)

½ pound rice hulls (optional but recommended)

1. Add the starch and the appropriate amount of water. (See below.)

Note: When commercial brewers perform a cereal mash, they add a small amount of malted barley (around 10 percent of the weight of the starch) and hold the mash at 150–160°F for around 30 minutes. This starts the enzymatic activity, which helps liquefy the

Starch	Gelatinization Temperature	Process Notes	Points per Pound per Gallon (PPG)
Rice	212°F	2 cups water per 1 cup rice; prepare the same as for eating (20 minutes for white rice, 60 minutes for brown rice).	1.033
Flaked Rice	N/A	Add straight to mash.	1.032
Corn Grits	212°F	2 cups water per 1 cup corn grits; prepare the same as for eating (boil and then simmer to reduce liquid).	1.033
Flaked Corn	158–167°F	Add straight to mash.	1.032
Flaked Oats	N/A	Add straight to mash.	1.028
Malted Oats	140°F	Add straight to mash.	1.026
Raw Wheat	125–129°F	Grind finely and cereal mash for 30 minutes before adding to mash.	1.026
Buckwheat	149–153°F	Cereal mash for 30 minutes.	1.028
Flaked Barley	N/A	Add straight to mash.	1.032
Spelt	135–156°F	Add malted spelt and spelt flour directly to the mash; raw spelt should be cereal mashed for 30 minutes.	1.032
Quinoa	135–156°F	Prepare the same as for eating, then add to mash.	1.028

mash and ensure that it's less likely to become gluey. For homebrewers working on a smaller scale, this is optional.

2. Cover the pot and then boil the mash for the recommended time, stirring occasionally to prevent scorching.

3. Proceed with your brew day as normal. Add the rice hulls to the mash along with the starch. (The rice hulls help prevent a stuck mash, but whenever you're using a large percentage of unmalted starch it's prudent to sparge slowly and gently.) If you do get a stuck mash, restir the mash, recirculate the wort until clear, and start sparging again.

OTHER STARCHES

Flours: I've experimented extensively with adding various flours into the mash. You might think this would create a gluey, stuck mash, but I've had no problems using up to 50 percent flour. I've had interesting results with tapioca flour and the water-chestnut flour from my local Asian market—they add distinctive flavors and change the mouthfeel in ways that make them similar to rice. Beers with lighter flavors are the best candidates for flours such as these, because they don't bring any color to the beer and the flavors are pretty subtle. They need a

clean platform in order to shine: American light lager, Kölsch, Munich helles, and cream ale are good choices. Avoid Belgian wit and other styles where the yeast and spicing will cover up the subtle flavors. As for process, I recommend mixing flours with the crushed malt before mashing in to make sure the flour doesn't clump. You can expect a similar PPG as you would have for malted barley. As with the other raw starches, use at least 20 percent in order to taste the ingredient, but don't go much beyond 40 percent unless you feel like risking a stuck mash.

Pumpkins and Winter Squash: Earlier and earlier each year, spiced pumpkin beers start to appear on store shelves. Most are generic amber ales with a cloying amount of pumpkin-pie spice added. More than likely, if the brewing company added any pumpkin at all, it was a tiny bit—more for marketing than taste contribution. That means it's up to homebrewers and craft brewers to make pumpkin beer worth drinking. The good news is that pumpkin and its many sister squashes can contribute a nice, subtle flavor to a beer while also providing a good amount of fermentable sugar. If you want to coax out some flavor, roasting fresh gourds is essential. Canned pumpkin puree won't taste the same. (See the next page.)

GLUTEN-FREE BEER

There are two ways to make gluten-free beer. One is to make a normal barley beer and add proprietary enzymes that theoretically will break down the gluten and make it safe to drink. This requires testing the beer, but on a homebrew scale these tests generally are unproven. They could make people sick, so I don't recommend them.

The only way to be absolutely sure of a gluten-free product is to use gluten-free ingredients. Since you don't have any enzymes to convert

starches and you don't have any husk material to keep the mash from sticking, you'll need something like brown rice or sorghum syrup, which are similar to malt extract but made from malted rice or sorghum. Sorghum is popular, but it tastes pretty cidery and winey, so the trick is to hide the flavor under yeast, spices, or hops. Emphasizing the hops or yeast in gluten-free beers also distracts from the missing barley flavor. See pages 168 and 171 for a couple of gluten-free recipes that take this approach.

Potato, Carrot, and Other Root Vegetables: The use of vegetables in beer is still largely uncharted territory. Other than the ubiquitous pumpkin beer or the rare sweet-potato beer, few veggies have joined the team. This is a chance for you to break new ground as a brewer. First roast almost any vegetable to develop caramelized tastes and then add it to the mash to convert any available starch to sugar. Start with 20 percent by weight in your grain bill and see where that takes you!

BREWING WITH PUMPKIN AND SQUASH

You can throw a few pounds of pumpkin, squash, or other gourd right in the mash, but it's best to roast them first to coax out their flavor. This takes time, but the payoff is a notable roasted-gourd taste that's pleasant and evocative of fall. I recommend around 3 pounds of gourd for a typical (1.050) 5-gallon batch.

If you want to accent the pumpkin with some traditional pumpkin-pie spice, that's up to you. But whatever you do, don't overspice! Remember that you can make a spice extract later (page 183), so go easy in the kettle. Don't add more than 1 teaspoon of any spice blend, whether pre-blended pumpkin-pie spice or your own custom blend, and add it at the end of the boil. Adding 1 teaspoon of vanilla extract at packaging can provide a nice complement to the spices as well.

YOU NEED

3–4 pounds small pie pumpkins or acorn squash

metal baking sheet

large plastic freezer bag

1 pound rice hulls (optional)

1. Preheat the oven to 350°F.

2. Cut the pumpkins or squashes in half and place them upside down on a baking sheet. Bake for 90 minutes. They should be soft and caramelized on the underside. If they aren't quite there after 90 minutes, give them more time in the oven.

3. Let the roasted pumpkins cool completely, then scoop out the flesh and put it in the freezer bag.

4. Squash the pulp in the freezer bag into as fine a mush as possible. You can keep the prepared pulp in the fridge for a few days before brewing if necessary.

5. On brew day, mix the squash pulp into the mash. Since pumpkin and squash often create a stuck mash, add the rice hulls to the mash. If it stops flowing during runoff, be patient and stir. It isn't unusual to have a 2-hour sparge if you use a large amount of squash.

SPELT FLOUR SAISON

This farmhouse-style ale uses spelt flour, but buckwheat, rye, or any other flour would work just as well. Mix the flour thoroughly with the crushed malt to prevent clumping when you mash in.

YOU NEED

basic brewing equipment (page 3)

9 gallons filtered brewing water (page 12)

9 pounds German or Belgian pilsner malt (75%)

3 pounds spelt flour (25%)

6 alpha acid units Centennial hops at 60 minutes (19 IBU)

1 Whirlfloc tablet

2 vials or packages French Saison WY 3711 yeast (or a 2-liter starter made from 1 pack, page 110)

5 ounces dextrose/corn sugar (optional, use only for bottling)

TARGETS

Yield: 5 gallons

OG: 1.062–1.064

FG: 1.007–1.009

IBU: 19

Note: A traditional Belgian mash allows for a thorough breakdown of proteins in the flour (122°F), the maximum amount of fermentability with a low saccharification rest (146°F), and a guarantee that everything gets gelatinized and converted by a short high-saccharification rest (162°F). It's a classic mash schedule when using unmalted grains or flours. If your system doesn't allow you to heat the mash, you can do a simple infusion mash at 149°F for 60 minutes. Your efficiency will drop, so you'll need to compensate by adding an additional pound of pilsner malt.

1. Mix the malt and flour with 4 gallons of water at 137°F or the appropriate temperature for the mash to stabilize at 122°F. Hold for 15 minutes, then apply heat and raise the mash to 146°F. Hold again for 45 minutes, then apply heat and raise the mash to 162°F. Hold a final time for 15 minutes.

Note: If you can't step mash, mix the malt with 4 gallons of water at 159°F or the appropriate temperature to mash at 149°F. Mash for 60 minutes.

2. Recirculate until the wort is fairly clear. Run off the wort into the kettle.

3. Sparge with 5 more gallons of water at 165°F. Run off the wort into the kettle.

4. Bring the wort to a boil. Boil it for 30 minutes. Add hops and continue to boil for 60 minutes. Add the Whirlfloc tablet at 30 minutes. Put your wort chiller into the wort at least 15 minutes before the end of the boil.

5. When the boil finishes, cover the pot with a lid or a new trash bag and chill to 70°F. Siphon the wort into your sanitized fermentor and pitch two packs of liquid yeast or a 2-liter starter.

6. During fermentation, allow the temperature to climb naturally. Don't worry if it approaches 90°F. Check the gravity after two weeks. When it falls below 1.010, you can package the beer.

7. Keg or bottle the beer. (If you're bottling, I recommend 5 ounces of dextrose/corn sugar for this beer.)

ATOMIC PUNK'IN ALE

This seasonal brew contains a dangerous amount of roasted pumpkin and requires a delicate hand with the spices. I like to think that I can taste the pumpkin, but it could be the pumpkin-pie spices tickling the synapses in my brain. The beer is a nice deep orange color with a good chewy mouthfeel, low hops, and just the faintest whiff of pumpkin pie baking a few houses away.

YOU NEED

basic brewing equipment (page 3)

10 gallons filtered brewing water (page 12)

11 pounds North American 6-row malt (for the extra husks)

1 pound crystal malt 60°L

3 pounds roasted pumpkin mashed in a freezer bag and added to the mash (page 163)

5 alpha acid units Willamette hops at 60 minutes (16.5 IBU)

1 Whirlfloc tablet

1 teaspoon pumpkin-pie spice blend at 5 minutes

2 vials or packages California Ale WLP001/ American Ale WY1056 yeast (or a 2-liter starter made from 1 pack, page 110)

2 teaspoons pumpkin-pie spice blend mixed with ¼ cup vodka for 1 week, strained through a coffee filter, and kept in a glass jar until bottling/kegging

3¾ ounces dextrose/corn sugar (optional, use only for bottling)

TARGETS
Yield: 5 gallons
OG: 1.060–1.066
FG: 1.013
IBU: 16.5

1. Mix the malt with 5 gallons of water at 171°F or the appropriate temperature to mash at 156°F. Mix in the mashed pumpkin mixture and mash for 60 minutes.

2. Slowly recirculate the wort until it's fairly clear. Run off the wort into the kettle.

3. Sparge with 5 more gallons of water at 165°F. Run off the wort into the kettle. If the runoff starts to slow down, you may have to restir the mash.

4. Bring the wort to a boil. Boil it for 15 minutes. Add the hops and continue to boil for 60 minutes. Add the Whirlfloc tablet at 30 minutes. Put your wort chiller into the wort at least 15 minutes before the end of the boil. Add the first spice addition at the 5-minute mark.

5. When the boil finishes, cover the pot with a lid or a new trash bag and chill to 65°F. Siphon the wort into your sanitized fermentor and pitch two packs of liquid yeast or a 2-liter starter.

6. Ferment at 65°F for one week, then let it warm to 70°F for another week.

7. Add the spice tincture to taste when you keg or bottle the beer. (If you're bottling, I recommend 3¾ ounces of dextrose/corn sugar for this beer.)

Note: When you're ready to keg or bottle, sample the beer and decide whether you want to give it more spice. If you do, slowly add the spice tincture until you reach the desired level of spiciness. If you're kegging, you can add more tincture at any time, so start with less than you think is right. After a pint or two, if you decide it needs more, throw in a couple of additional tablespoons.

GLUTEN-FREE IPA

This beer lacks malt flavor because there's no malt in it, but its intense hop character makes up for the deficiency. It has huge hop aroma and taste, with a crisp bitterness. This is a 3-gallon recipe.

YOU NEED
basic brewing equipment (page 3)

4 gallons filtered brewing water (page 12)

6 pounds Briess White Sorghum Syrup

10½ alpha acid units Columbus hops at 60 minutes (48 IBU)

2.7 alpha acid units (about ⅓ ounce) Amarillo hops at 30 minutes (10 IBU)

3.7 alpha acid units (about ⅓ ounce) Centennial hops at 15 minutes (7.6 IBU)

3.7 alpha acid units (about ⅓ ounce) Simcoe hops at 5 minutes (4 IBU)

⅓ ounce Amarillo, Simcoe, and Centennial hops at end of boil

1 Whirlfloc tablet

1 vial or package California Ale WLP001/ American Ale WY1056 yeast

⅔ ounce Simcoe, Columbus, and Cascade hops (dry hop)

2½ ounces dextrose/corn sugar (optional, use only for bottling)

TARGETS
Yield: 3 gallons

OG: 1.066

FG: 1.010

IBU: 70

1. Bring 4 gallons of water to a boil and dissolve the sorghum syrup in it.

2. Add the first addition of hops and continue to boil for 60 minutes, adding the other hop additions as indicated. Add the Whirlfloc tablet for the last 30 minutes. Put your wort chiller into the wort at least 15 minutes before the end of the boil.

3. When the boil finishes, cover the pot with a lid or a new trash bag and chill to 65°F. Siphon the wort into your sanitized fermentor and pitch the yeast.

4. Ferment at 65°F for one week, then let it warm to 70°F for another week. Add the dry hops after primary fermentation slows down (usually after five to seven days).

5. Keg or bottle the beer. (If you're bottling, I recommend 2½ ounces of dextrose/corn sugar for this beer.)

ST. HUBBINS DUBBEL

The combination of Belgian ale yeast, Belgian candy syrup,
and coriander creates a lot of complexity in this gluten-free brew.
Note that this is a 3-gallon recipe.

YOU NEED

basic brewing equipment (page 3)

4 gallons filtered brewing water (page 12)

6 pounds Briess White Sorghum Syrup

1 pound Dark Candi Syrup (D2)

4½ alpha acid units Fuggle hops at
    60 minutes (29 IBU)

1 Whirlfloc tablet

¼ ounce ground coriander at 5 minutes

1 vial or package Trappist Ale WLP500/
    Belgian Ale yeast WY1214

2½ ounces dextrose/corn sugar
    (optional, use only for bottling)

TARGETS

Yield: 3 gallons

OG: 1.073

FG: 1.011

IBU: 29

1. Bring 4 gallons of water to a boil and dissolve the sorghum syrup and Candi syrup in it.

2. Add the hops and continue to boil for 60 minutes. Add the Whirlfloc tablet for the last 30 minutes. Put your wort chiller into the wort at least 15 minutes before the end of the boil. Add the ground coriander during the last 5 minutes of the boil.

3. When the boil finishes, cover the pot with a lid or a new trash bag and chill to 65°F. Siphon the wort into your sanitized fermentor and pitch the yeast.

4. Ferment at 65°F for two days, then let the temperature slowly rise to 72°F and hold it there for ten days.

5. Keg or bottle the beer. (If you're bottling, I recommend 2½ ounces of dextrose/corn sugar for this beer.)

SUGARS
AND
FRUIT

SUGARS

Sugars used to have a bad reputation in the homebrew community, but using sugar in moderate amounts (5–15 percent) can be good for certain recipes. Simple corn sugar ferments completely, making strong beers drier and more drinkable. Many other sugars have complex flavors in their own right that can be wonderful in the right beer—and fun to brew with.

Unlike grains and other starches, sugars don't need to be mashed, gelatinized, or malted. Just toss them into the boil (or in some cases, the fermentor itself) and you're done. Most sugars are more fermentable than malted grain (except for molasses, typically only 50 percent fermentable), so higher attenuation and less body in the finished beer are the norm. It's safe to assume that you'll get 42–45 gravity points from a pound of sugar, no matter the variety. That means a 5-gallon batch would have a gravity increase of slightly more than 0.008 for every pound of sugar.

To preserve its delicate aroma, any flavorful sugar should be added at the end of the boil, once you turn off the heat. For maximum effect, you can add the sugar to the fermentor after primary fermentation completes (page 174).

Corn, Cane, and Beet Sugars: Highly refined sugars, such as corn sugar, have no real tastes of their own. They just add a high amount of alcohol by weight. The higher percentage you use, the thinner the body of the finished beer. Sounds bad at first, right? But some beers really can benefit from this result. Many brewers use sugar to make a big beer crisper and easier to drink. Consider a Belgian tripel: It's normal for this beer to have pure, tasteless sugar as part of its grain bill. Be careful, though—too much sugar will give your beer a winey note because of the extra alcohol. My rule of thumb is to use no more than 15 percent, with 5–10 percent a more common range.

Brown Sugars: A variety of darker sugars can add tastes of rum, toffee, or molasses to beer in addition to kicking up the alcohol in the same way that corn sugar does. All brown sugars aren't created equal, though. Standard, cheap brown sugar is usually just cane sugar with a tiny bit of molasses added for color; you want to use less-refined brown sugars such as turbinado or Indonesian

gula jawa for delicious, complex flavors. They work well in brown ales, Belgian dark ales, British old ales, and many other styles.

South and Central American Sugars: These brown sugars often look quite rustic and sometimes come wrapped in cornhusks. Basically, they're pure sugarcane juice that's been evaporated over a fire. The less refined they are, the better they taste. Brazillian rapadura and Mexican panela, or piloncillo (formed into little cones), all have complex caramel and toffee tastes and are fantastic in any beer that can handle an extra point or two of alcohol.

Palm Sugar: This tan-colored sugar usually is molded into small disks. It has a subtle maple-syrup taste and is one of my favorite sugars for brewing, especially in British ales. Look for it at your local Asian supermarket.

Belgian Candi Syrup: This form of sugar is used throughout Belgium to lighten the body of their strong beers, often adding flavor at the same time. For years, homebrew shops sold bags of candy-sugar rocks, and homebrewers assumed that was what the Belgians were using. Now we know that they use a concentrated sugar syrup that comes in a variety of flavors, depending on the caramelization of the sugar. Luckily, several suppliers offer these syrups, so you should be able to find them at your homebrew store or online. I recommend the darkest version available for Belgian dark ales. For Belgian golden ales and tripels, the flavor contribution is truly minimal. Save your money and stick with cane sugar.

Honey: This sticky substance has been used to make alcohol since . . . well, forever. In mead, all of the fermentable sugar comes from honey. Braggot, the love-child of mead and beer, contains a 50-50 blend of malt and honey. Beer brewers have something of a love-hate relationship with it. A honey wheat awakens feelings similar to those aroused by a raspberry wheat: It's not inherently bad, but so many brewers have done it badly.

Honey can blend wonderfully into beer recipes, though, and not just the ones in which it's the star of the show. Honey ferments out completely, so it won't add any

ADDING SUGAR LATE

When you add spices to a beer, you lose more delicate aromatics the longer you boil them. The same goes for adding sugars: They don't contain as many volatile compounds as spices do, but they have complexities of flavor that you don't want to lose. Higher wort gravities will inhibit hop isomerization, so you shouldn't add sugars before the last 15 minutes of the boil, even if you care little about their flavor.

If you're reading this, you're likely using a specialty sugar, and you probably want to preserve its character. Very delicate sugars, such as maple syrup or honey, should be added after the heat has been turned off if you want the most flavor. Some brewers go the extra mile and add the honey or maple syrup into the fermentor after the primary fermentation completes. If you want to go this route, dissolve the sugar in enough water to absorb it and heat it to a boil. Then cover the pot and let it cool a bit before adding it to the fermentor.

It may be tempting to take this logic as far as possible and use unusual sugars to prime a beer in the bottle or keg, but that's risky business. Each type of sugar has its own level of fermentability, ranging from less than 50 percent for molasses to 100 percent for cane sugar. Substituting one for the other can result in under-carbonated beer, bottle bombs, or a foamy keg. When it comes to carbonation, stick with tried-and-true techniques.

sweetness to the beer itself. Many brewers recommend adding honey as late as possible—either at the end of the boil or after primary fermentation completes—to retain more of the volatile aromatics. (See left.) If you're trying to get the taste of honey to come through in your brew, remember that, in addition to when you add it to your beer, the source of the honey can make a big difference, too. Experiment with different varieties, from bright, citrusy orange blossom to dark, licorice-flavored buckwheat honey.

Molasses: This is strong stuff and can easily overpower a beer, so use $\frac{1}{2}$ cup or less per 5 gallons no matter how strong your beer. Limit its use to dark beers that will let the flavor meld with the malt without overpowering it. Three common grades are available in America. The lightest is mild (or Barbados), the middle grade is dark (or second), and the darkest is blackstrap. No matter what grade of molasses you buy, make sure to get the unsulfured variety. Molasses made from younger sugarcane has sulfur dioxide added as a preservative, which will add an off-flavor to your beer. Always taste a spoonful before you add it. If it has any metallic flavors, don't use it. British black treacle is a close relative of molasses that's slightly lighter in taste. It also comes in a light version called "golden syrup." It has a nice taste but is somewhat difficult to find and can be expensive when purchased stateside. Add molasses during the last five minutes of the boil to volatilize any unpleasant aromas while keeping most of the complexities.

Maple Syrup: This is a classic American ingredient. Native Americans first made it, and colonial settlers embraced it. Various species of maple trees, including sugar maple, red maple, and black maple, are traditional varieties, although any maple tree will yield syrup. The trees exude a sugary sap in the spring. To access the sap, multiple holes are bored into the trunk of a maple tree, and the substance is collected in a bucket. The liquid is only mildly sugary and needs to be boiled down to create syrup. It takes 20 to 50 gallons of sap to make just 1 gallon of syrup!

Maple syrup generally is labeled as Grade A or Grade B, with Grade A being lighter in color and Grade B being darker and having a more intense maple flavor. As a brewer, you want Grade B because it has better maple aromatics. Real maple syrup is very expensive, but do I really have to tell you not to use fake maple syrup? Save the cloying substitute for brunch. To maximize taste contribution, add maple syrup as late in the fermentation as possible to prevent the delicate aromas from escaping the airlock.

FRESH FRUIT

Fake fruit–flavored beers have a tainted history. The brewing of real fruit beers, however—from Belgian lambic producers to small craft brewers who make stellar summer beers with ripe fruit—has a distinguished history. No matter what style of beer you're making, if you're brewing with fresh fruit for the first time, your biggest surprise probably will be how the fruit's water content affects the beer. Many homebrewers think of fruit as sugary—and thus an ingredient that will add alcohol to a beer the same way sugars do. It's true that fruit is sweet, but the sugar generally is more than offset by the water content. In other words, the sugar-to-water ratio usually means the alcohol percentage of the beer remains the same even as the pounds of fermentables increase. So your batch becomes bigger in size but not stronger.

Here's a list of common fruits and some starting points for adding them to a typical 5-gallon batch of beer. The orchard is yours for the taking! Almost any fruit is fair game. Just keep in mind that, for some fruits, removing residue can prove problematic. Let's just say that there's a reason—besides taste—that not many people make banana-flavored beer.

Apricots and Peaches: These stone fruits are some of the best to add to a blonde-colored beer. If you have the choice, apricots are the way to go. A Belgian blonde, witbier, or even a Berliner weisse tastes great with stone fruit. About 10–12 pounds of the ripest fruit you can find is a good starting point for a 5-gallon batch. When using fresh peaches, it's essential to use perfectly ripe or even overripe fruit.

Blueberries: These have a particularly high water content, so you need a lot of them (12–20 pounds) if you want the taste to come through. You won't end up with a blue beer. The finished product will have a reddish hue.

Blackberries: These berries are more aggressive and tart than blueberries and can handle a more assertive beer style. Use 10–15 pounds per 5 gallons.

Cherries: Belgian lambic brewing uses these widely. Their flavor also blends well with American-style ambers or dark beers, which is why cherry stouts and sour browns are popular. A good starting point is 10–15 pounds for a 5-gallon batch. It's a good idea to stay away from all so-called natural flavorings, but that rule applies *doubly* to cherry. It will make your beer taste like cough syrup.

Raspberries: Like cherries, raspberries are a classic choice in traditional Belgian brewing. I like using them because their flavor doesn't get lost in a beer like so many other fruits. Since they're so intense, you don't need to use a huge amount of them. As few as 5 pounds make a nice fruit beer, though you can go up to 10 pounds.

Tropical Fruits: Many fruits haven't been experimented with widely in the commercial world, but that doesn't mean they won't work if you throw them into a beer. Mango, passion fruit, pineapple, and kiwi are all interesting candidates for experimentation. My rule of thumb for experimental fruit is to use 2 pounds per gallon. Also, it's worth noting that some of these tropical fruits, such as passion fruit and pineapple, come in juice form as well (page 178).

Citrus Zest: Toss aside those chunks of dried Curaçao orange peel—the only flavor you'll get from them is unpleasant bitterness. If you want citrus character in your Belgian beers, give fresh citrus peel a try. Let an individual fruit shine and make a tangerine witbier, or use a variety of citrus fruit in combination (tangerines, grapefruits, oranges, and lemons). Zest three or four pieces of fruit, making sure not to overzest and dig into the bitter pith, and add the zest at flameout. You can also add the zest into the fermentation vessel or straight to a keg, but use about half as much if you go that route. If you're concerned about the minor sanitation risk, steep the zest in a bit of vodka.

TOO MUCH WATER, TOO LITTLE TASTE

Just because you love the taste of a certain fruit doesn't mean it'll work in a beer. Sometimes the juiciest fruits get lost in translation. For example, fresh strawberries sound like they'd make a great beer, don't they? Unfortunately, their flavor is so delicate that you'd have to use huge amounts (something like 20 pounds) for the beer to taste the way you want—and the aroma and taste would fade within a few weeks. Watermelon also sounds ideal for a summer sipper, but, again, the water content is so high and the taste so delicate that it just doesn't come through.

Surprisingly, apples fall into this camp, too, because their taste is so mild that it's tough to get them to make an impact as whole fruits. (They also can add a cidery flavor to your beer that's similar to a classic off-flavor considered a defect in most beers.) The more assertive a fruit, the better it is in beer.

HOW TO PREPARE AND USE FRESH FRUIT

The dilemma for brewers is whether and how to sanitize fresh fruit. Lambic brewers just toss the unsanitized fruit in their beer, pits and all. But for most brewers, the idea of inviting trouble from the wild yeasts on the fruit is pretty scary! Which brings us to the options for disinfection: heat, cold, or both. (Star San isn't a possibility here.)

A good practice for using any fruit is to freeze it for a day or two, which will break up the flesh and allow for maximum flavor extraction later. Then you can take the thawed fruit and add it straight to your beer after primary fermentation. The cold won't kill all of the wild stuff, but most brewers assume the brewing yeast will overcome any stray bad guys. You also can heat the thawed fruit to 160°F and hold it for 10 minutes before adding it to your beer if you want some extra peace of mind. If that sounds like too much work, try using pre-sterilized commercial fruit purees. They lack the romance of locally picked fruit, but they're the go-to source for consistent, no-worry fruit.

Belgian brewers often leave fruit in their beer for four months, but that's excessive for a typical fruit beer. At around two weeks, you've extracted almost all of the fruit character, especially if using a puree. Once you go more than four weeks, you could start losing fruit character. Before kegging or bottling, chill the beer below 40°F and then add a fining such as gelatin (page 123) to compact the fruit pulp and make it easier to transfer the beer to a keg or bottling bucket.

OTHER FORMS OF FRUIT

Dried fruit, purees, and fruit juices can be used in brewing. Unlike fresh fruit, however, dried fruits (and many purees) can add extra alcohol to the beer since they aren't diluting it with a bunch of water. Sanitizing dried fruit is difficult. You could soak the fruit in booze and add them into the fermenter when fermentation ends. Another option is to heat the dried fruit in a small amount of water to 160°F to pasteurize it, then add it to the fermenter.

Dried Cranberries, Cherries, and Grapes (Raisins): These popular snack foods are all good candidates for adding to an appropriate beer. What style of beer is appropriate? Ideally, one that can handle a good dose of acidity. A dark, roasty stout would work, as would a light wheat beer. A heavily hopped IPA probably wouldn't work well.

Most dried fruits have a raisiny aspect well suited to beers that showcase darker crystal malts with a similar flavor (such as Special B, page 32). Belgian dark ales often are described as raisiny. Dried fruit can be added at any point in the brewing process: mash, boil, or fermentor. As little as ½ pound of dried fruit can add a significant amount of character and is my recommended starting point.

Dried Citrus Peel: Although dried peel is inferior to fresh peel, you certainly can use it when the option to zest something fresh doesn't exist. Dried lemon peel and sweet orange peel are killer in any Belgian witbier, saison, or blonde. Start by adding around ½ ounce during the last few minutes of the boil to allow the dried fruit to rehydrate and release its goodness.

Fruit Purees: When fresh fruit is out of season, fruit purees come to the rescue. They can be pricey, but the appeal of pre-sanitized fruit with no stems or seeds is hard to resist. Oregon Fruit Products is the largest producer of purees, and they offer their fruit in everything from 2-pound cans to 22-pound packs. They're used widely by commercial breweries and should be your go-to fruit source any time local fruit is less than perfect. You usually can use less of them than fresh fruit for the same result. I typically use around 25 percent less when using fruit purees in a recipe that calls for fresh fruit.

Fruit Juices: You might lose a small amount of aroma or taste from the additional processing, but it's hard to dismiss the ease of using a sterilized, filtered juice. There's no pulp to rack off of, and sanitizing the fruit isn't a concern, so what's not to like? The trouble is finding a juice not loaded with sugar or preservatives, such as potassium sorbate, which prevent your yeast from working. Several grocery-store brands are 100 percent juice with no preservatives, but you need to make sure they aren't diluted with other juices such as apple or cranberry. Natural-food stores and international markets are a little more expensive, but they often have smaller bottles of pure concentrated fruit juices such as cherry, cranberry, and tropical fruit. Once you find a juice you like, start with 2 quarts in a 5-gallon batch of something like a blonde ale or witbier, adding the juice after fermentation slows. Give the yeast at least three to four days to consume the sugars.

Type Of Fruit	Suggested Quantity for 5-Gallon Batch	When to Add	Color Contribution	Recommended Beers
Apricots and peaches	10–12 pounds	End of boil or secondary	Low	American wheat, lambic
Blueberries	12–20 pounds	End of boil or secondary	Medium	Delicate beers, but can work in the right stout or porter
Blackberries	10–15 pounds	End of boil or secondary	High	Delicate beers, but can work in the right stout or porter
Cherries	10–15 pounds	End of boil or secondary	Medium	Imperial stout, dubbel, lambic
Raspberries	5–10 pounds	End of boil or secondary	High	Anything from light wheat beers to imperial stout
Tropical fruit	Varies	End of boil or secondary	Low	Lighter beers
Citrus zest (fresh)	Zest from 3–4 pieces of fruit	End of boil or secondary	None	Lighter beers
Citrus zest (dried)	Zest from 2–3 pieces of fruit	End of boil or secondary	None	Lighter beers
Raisins, prunes, dried cherries	1/2 pound	End of boil or secondary	Low	Dark-brown or ruby-red beers
Fruit purees	Varies based on fruit	Secondary	Varies	Varies based on fruit
Fruit juices	Varies	Secondary	Varies	Varies based on fruit

BASIC BLONDE ALE

This tasty golden ale is similar to a Kölsch. It drinks fine by itself, but it's designed as a base beer that can showcase a specialty ingredient such as an herb or a fruit. The mix of pilsner and 2-row malt gives it a neutral malt profile that lets the specialty ingredient shine. The recipe calls for Hallertau hops, but you easily can substitute any low to medium alpha-acid hop (3–6 percent alpha acid).

YOU NEED

basic brewing equipment (page 3)

8 gallons filtered brewing water (page 12)

4 pounds German pilsner malt (44%)

4 pounds North American 2-row malt (44%)

1 pound wheat malt (11%)

3½ alpha acid units German Hallertau hops at 60 minutes (22 IBU)

1 Whirlfloc tablet

2 vials or packages California Ale WLP001/ American Ale WY1056 yeast (or a 2-liter starter made from 1 pack, page 110)

5 ounces dextrose or corn sugar (optional, use only for bottling)

TARGETS
Yield: 5 gallons
OG: 1.047–1.049
FG: 1.010–1.012
IBU: 22

1. Mix the malt with 3 gallons of water at 165°F or the appropriate temperature to mash at 150°F. Mash for 60 minutes.

2. Recirculate the wort until it's fairly clear. Run off the wort into the kettle.

3. Sparge with 5 more gallons of water at 165°F. Run off the wort into the kettle.

4. Bring the wort to a boil. Boil it for 30 minutes. Add the hops and continue to boil for 60 minutes. Add the Whirlfloc tablet at 30 minutes. Put your wort chiller into the wort at least 15 minutes before the end of the boil.

5. When the boil finishes, cover the pot with a lid or a new trash bag and chill to 65°F. Siphon the wort into your sanitized fermentor and pitch two packs of liquid yeast or a 2-liter starter.

6. Ferment at 65°F for one week. During the second week, let it warm to 68–70°F and add a specialty ingredient (see the next page) once primary fermentation slows down, usually after five to seven days.

7. Keg or bottle the beer. (If you're bottling, I recommend 5 ounces of dextrose/corn sugar for this beer.)

VARIATIONS

Fresh Herbs: Many herbs can add a pleasant aroma to this beer. My favorites are Thai basil, lemon basil, lemongrass, and lemon verbena, and 1 ounce is a good starting point. Soak the herbs in vodka or boiling water for a few seconds and add them to the fermentor after 1 week of fermentation. Wait another week before bottling or kegging.

Fresh Fruit: Other great options for this beer include the addition of raspberries (5 pounds) or blueberries (10 pounds). Freeze the fruit overnight to help break down the pulp, then add it to the fermentor after 1 week of fermentation. You can boil the fruit first if you want, but the alcohol in the beer and the low pH should keep any wild yeast from taking hold. Let the fruit sit in the fermentor for 7–10 days before bottling or kegging.

HERBS, SPICES, AND OTHER INGREDIENTS

HERBS AND SPICES

Herbs have a long history in brewing. For centuries before hops first were added to beer, there was gruit beer, an ale that was flavored with an herb mixture known as gruit. The holy trinity in all gruit mixtures consisted of sweet gale (aka bog myrtle), yarrow, and wild rosemary. Other herbs were added according to local tastes or availability, including mugwort, juniper, wormwood, heather, and sage. Some of these herbs have sedative or medicinal properties, which can be potentiated by the alcohol. Others, such as wild rosemary, can be dangerous, so do your research! For additional information, check out *Sacred Herbal and Healing Beers* by Stephen Harrod Buhner.

Use your senses and preferences to determine whether you actually want an herb or spice in your beer. Crush some between your palms or in a blender and smell it. Chew on a pinch. Think about how that character will blend into a beer. Imagine how it will blend with hops or dark malts. Dunk a sprig of a particular herb or otherwise spike a pint for a quick research and development project.

You can add herbs at knockout, to secondary, or as a tea or tincture at bottling (page 184). Each herb is different, but for the most commonly used fresh herbs (such as basil and lemon balm) add 1 ounce per 5-gallon batch for subtle character, or up to 2 or 3 ounces per 5 gallons for a more intense flavor. For dried herbs, the equivalent is about 1/8 to 1/4 ounce per 5-gallon batch. If you want to add fresh herbs after primary fermentation, dunk the herbs into boiling water for a second or two, then add them to the fermentor. Find the freshest dried herbs and spices available, because they can lose their aromatics quickly.

Basil: Basil comes in many varieties, each with its own unique flavor, such as lemon, pineapple, and Thai. The only variety I wouldn't use is typical Italian basil—you don't want drinkers thinking about spaghetti sauce when smelling your beer! Basil is an excellent addition to many beers, including Belgian blonde, saison, and American wheat.

Lemon Verbena: This herb has an intense lemon aroma, which can be nice on its own or in combination with basil. It's a good candidate for a knockout addition, because it doesn't carry as large a risk of overpowering your beer as do other herbs.

Wormwood: Wormwood is very bitter and can be used to bitter beer, as it does the infamous liquor absinthe. The bitterness is coarse and strong, though, so make a tea and add it to taste rather than adding it to the boil (see page 184).

Lemon Balm: Lemon balm lacks the pure lemon flavor of lemon verbena. However, it has a complex earthiness that can play nicely with some beers.

Bog Myrtle: Also called "sweet gale," bog myrtle has a mild, earthy, dried-grass character. It's essential in a gruit ale.

Cilantro: I haven't used cilantro as much as other herbs, but I've seen it here and there for "tap-room only" and limited-edition beers. Some people think it smells like soap, so you're better off blending it with other herbs.

Chamomile: Chamomile often is used as a secret third ingredient in Belgian witbiers (in addition to coriander and orange peel). A little goes a long way, so start with 2 tablespoons and go from there.

Rosemary: Rosemary is resinous and intense. As a result, it blends well with dark malts and piney hops. Because it can take over a beer quickly, boil up a tea and add it to taste at bottling or kegging. Around 1 to 2 ounces of fresh rosemary should be about right, but it really depends on the base beer and its hops.

Mint: Mint can be tricky to work into a beer. If you balance it perfectly, a bright mint can blend well with chocolate malts. If you overdo it even a little, the result can taste like toothpaste. Either way, it's difficult to drink a 5-gallon keg of mint beer. Trust me.

Juniper Berries: These berries have a piney, resiny taste that blends well with aggressive hopping in an IPA or double IPA. Crush the berries before using them and start with 1 to 2 ounces.

Ginger: Ginger is another good candidate for a secret ingredient in witbiers, or you can try it on its own in an American wheat. Use 4 ounces of chopped fresh ginger in 5 gallons for a subtle kick or up to 1 pound of ginger for a stronger flavor. Ginger works well when added during the last 15 minutes of the boil.

Grains of Paradise: These are like small black peppercorns, but they have a citrusy note as well. Most recipes call for just a pinch, but I found it undetectable until at least ¼ ounce, and still subtle at ½ ounce. Crush it and add during the last 5 to 10 minutes of the boil.

Caraway Seeds: When people think of caraway seeds, they think of rye bread. Try adding 1 tablespoon of crushed caraway seeds to a dark amber ale and calling it a rye beer—in the same spirit as adding pumpkin-pie spice to a beer with no pumpkin in it. Even if there's no rye in the beer, people will think of rye.

Mugwort: Mugwort has a distinctive aroma that elicits the same knee-jerk "I hate it" reaction as cilantro for some people. I've had to dump an entire keg of beer made with mugwort.

Sage: Sage adds an earthy dankness that blends well into porters and stouts. It works as a tea or tincture, but you can try adding 1 ounce of fresh sage at the end of the boil.

MAKING A TEA OR TINCTURE

The rule of thumb that every commercial brewer follows is "less is better." But sometimes there's no way to know that you've gone too far until you've gone too far. The good news is that non-fermentable seasonings such as herbs and spices are easy to dial in without wasting more than a bottle or two of beer since they can be added to the finished beer to taste. The two most common ways to do this are by making a tea or a tincture.

➡ **To make a basic tea:** Combine 2 to 3 ounces of fresh herbs or ¼ ounce of dried herbs with 2 cups of boiling water and then allow them to steep, covered, with the heat turned off until cool. Strain through a coffee filter into a sanitized jar. This makes enough for a 5-gallon batch, so it's quite concentrated and perfect for adding a little at a time.

➡ **To make a basic tincture:** Combine 2 to 3 ounces of fresh herbs or ¼ ounce of dried herbs (more common for tinctures) with enough vodka to submerge them. Let the mixture steep for a few days, then strain it through a coffee filter into a sanitized container.

No matter which method you choose, you now have a potent extract that you can add to taste for the perfect balance of tastes. If you keg your beer, add a small amount of tea/tincture, shake the keg, and pour yourself a sample. Repeat until the level of the spice or herb is to your liking. If you're bottling, pour 12 ounces of beer from your bottling bucket into a glass and, using a milliliter dropper, keep adding the tea/tincture to the glass until the proper level is reached. Then multiply the milliliters you added by 48 and that will be the proper amount to add for a full 5-gallon batch.

OTHER INGREDIENTS

"There are no rules, there are no rules." Keep repeating that to yourself as you think about ingredients to add to your beer. On the other hand, be realistic. You're probably going to dump some beer if it doesn't come out like you expected. That's part of being an experimental brewer, and you shouldn't be afraid! But if you're out of beer, maybe you should fill the fridge with a pale ale before you attempt a *Brettanomyces guanabana* coffee beer.

Coffee

It's funny how specific some brewers are when it comes to their ingredients, yet they still think of coffee as an interchangeable product. Coffee beans, like wine grapes, can vary wildly in taste, and you should consider a few things when adding coffee to beer.

➡ **Origin:** Coffee beans from different parts of the world have different flavors. Certain regions, such as Latin America, are known for a brighter, fruitier taste, whereas others are spicier or earthier. If you're already a coffee lover (and if you're adding coffee to your beer, you probably love coffee), you already know what I mean. There's no reason to avoid the coffees with higher perceived acidity if you like their taste, but, to dampen some of the effect, cold brew them instead of adding them at flameout (see below).

➡️ **Roast:** From light brown to dark black and oily, roast makes a difference. Dark-roasted beans have a lot more roast flavor than more mildly roasted beans. If that's what you want, go for the French roast. If you want more coffee aroma than roast aroma, stick to medium or light roast.

➡️ **Grind:** Always use a grind appropriate for your extraction method. If you grind too fine or too coarse, you won't get the desired taste.

➡️ **Extraction/Addition Method:** Cold brew a batch of coffee and add it to taste at bottling or kegging. Cold brewing is brewing coffee at room temperature for an extended period of time instead of at high temperatures for a short period of time. It changes the flavor and quality of the extraction in a way that lets you keep some of the coffee essence while minimizing perceived acidity. Also, cold-brewed coffee can last in the fridge for weeks as a concentrate, as opposed to hot coffee, which—as most of us know—is no good the next day.

To cold brew, add 6 ounces of coarsely ground coffee (the same grind as for a French press) per quart (4 cups) of water. Let it sit between 12 and 24 hours, then pour it through a sanitized strainer or cheesecloth into another sanitized container. Add to taste at bottling or kegging. Pour in half the coffee, taste it, and go from there.

➡️ **Dry beaning:** My favorite method of adding coffee is also the easiest: tossing whole beans into the fermenter as you would dry hops. Start with 2 ounces of whole beans. As few as 48 hours of contact time are all you need to extract most of the coffee character from the beans. Add more beans after several days if you want a stronger coffee taste.

➡️ **Quantity:** Start with 6 ounces for cold brew or 2 to 3 ounces for a flameout addition. If you want more coffee taste, you certainly can use more coffee.

It sounds like a lot to consider, but you probably already have a coffee in mind. Think about the flavor and how it will work with your beer as you drink the coffee. Pay close attention to the aroma, the most notable part of the coffee once it's blended with the beer.

If you don't have a coffee in mind, talk to someone at your favorite independent coffee shop and ask for descriptions of their different beans. Tell them what aromas and tastes you want, and let them help you pick a coffee! A big, roasty stout probably would benefit from a darker French roast, while a caramely red ale would be better with the mild, berry flavors of a lighter roast. You probably want the coffee to be at a threshold similar to that of herbs: notable but not overpowering.

Cocoa and Chocolate

That maltsters make a malt that tastes like chocolate shows that chocolate works well in beer—dark beer in particular. But before you start chucking a dozen of your favorite chocolate bars into the boil, you need to know that the fat in regular chocolate will form little balls of fat in your beer, and nobody likes chunky beer. So you need to use chocolate that has minimal fat.

Low-fat chocolate sounds like an oxymoron, but cocoa powder and cacao nibs both fit the bill. Cocoa powder can be very high quality, but it tends to make a gummy sediment at the bottom of the fermentor that's hard to rack off. That means cacao nibs are the best option.

I've seen recommendations for allowing *weeks* of contact time for cacao nibs, but as few as 12 hours will pick up the majority of the taste, and you need no more than two days for full absorption. Since cacao nibs (and cocoa powder) have been roasted, there's no need to

worry about sanitizing them before adding them to a secondary fermentor. About 2 to 3 ounces of nibs (or cocoa powder) in a 5-gallon batch should give you a notable—but not overpowering—chocolate aroma and taste. The nibs don't have to be ground; just toss them in. If you live near a chocolate maker, ask them what varieties of nibs they use. Like coffee, cacao nibs vary quite a bit depending on their origin and the amount of roasting. Brew a 10-gallon batch, splitting it into two 5-gallon fermentors at secondary, and add two different types of nibs for a fun side-by-side tasting.

Vanilla

Vanilla is a classic in holiday beers. Like many herbs and spices, fresh is best, which means using vanilla beans if you can. Add 2 to 3 vanilla beans split down the middle at the end of the boil. Then transfer them with the beer to the fermentor. (Don't strain them out.) You also can add vanilla beans to the secondary or the keg instead, but, since they're a fermented bean pod, sanitize them first.

An easier approach is to use vanilla extract (the pure stuff, not that artificial junk), and add it to taste at bottling or kegging. One or two teaspoons is usually enough. Vanilla can make for an excellent accent flavor in a beer. Adding just one or two beans to a beer to which you're also adding chocolate or coffee can meld the flavors and create a whole more interesting than the parts.

Spirits

Most drinkers prefer their liquor in shot glasses beside their pints, but it was only a matter of time before some guy dropped his shot into his beer—and not just as a bomb or depth charge. Commercial brewers can't add alcohol to a beer, so it's up to the homebrewers to chart this territory. For straight spirits with no residual sugar, such as whiskey and gin, there's no need to factor in sugar that would ferment out. Since there's no risk of further fermentation, these alcohols can be added not only to the primary but also at packaging. What about a strong scotch ale with ½ ounce of single malt scotch added to each bottle?

Other liquors have interesting tastes and a substantial sugar content. That sugar will ferment out, which means these spirits should be added only during primary fermentation. Try a Trappist quad spiked with Bénédictine, or a Jägermeister-infused doppelbock! As for how much to add, try it on a commercial beer first. Using a dropper with milliliter markings, add a small amount of the liquor you want to use to a commercial beer similar in style to the one you're going to brew (keeping in mind that the yeast will eat the sugar). When you get to an amount that tastes right, scale up for a 5-gallon batch.

Hot Peppers

The bright, herbal character of mild Anaheim or New Mexican chiles can be fantastic when used sparingly in a Munich helles or an American wheat. Just the right amount of the smoked jalapeños known as chipotles can give a beer a subtle hint of smoke and a mild warming effect. But chile beers also can be horrible firebombs that destroy your taste buds for the rest of the night. Finding the right balance is tough.

Since every chile is different—even chiles off the same plant—you need to taste each one. If a chile is blistering hot, you won't be able to get the taste of the chile into the beer without it becoming too spicy to drink. If you think it's mild enough or close to it, remove the seeds and the veins holding the seeds. They don't contribute any flavor, and you want to add as much chile taste as you can before the heat builds to all-out spicy. You can reduce the impact of the seeds further by rinsing the chile after you deseed it. Many restaurants do this for jalapeño poppers.

Another way to reduce the spiciness of peppers is to roast them whole over an open flame until blackened, and then remove the skin and seeds. As you'd expect, this will change the flavor of the chiles. When roasted, they take on a smoky, sweet taste as opposed to the bright, fresh taste of raw chiles.

Lighter beers are better suited to fresh chiles, and darker beers are a better match for roasted or smoked chiles.

NEW MEX PILSNER

As guest brewers, my wife and I made this beer at the Abita Brewing Company in New Orleans back in the early 1990s. We called it "Nacho Mama," and it was a big hit. It drinks like a traditional German pilsner, but you get a bright hit of fresh New Mexican chiles instead of the expected Hallertau or Saaz hop aroma. The heat is minimal, which keeps the beer very drinkable and makes it a fantastic brew to bring to a Cinco de Mayo party!.

YOU NEED

basic brewing equipment (page 3)

8½ gallons filtered brewing water (page 12)

9 pounds German Pils malt (100%)

7 alpha acid units German Perle hops at 75 minutes (31 IBU)

1 Whirlfloc tablet

10 fresh, green New Mexican chiles, washed, stemmed, and seeded, at 5 minutes

3 vials or packages German Lager WLP838/ Bavarian Lager WY2206 yeast (or a 2-liter starter made from 1 pack, page 110)

4.6 ounces dextrose/corn sugar (optional, use only for bottling)

TARGETS
Yield: 5 gallons
OG: 1.048–1.050
FG: 1.010
IBU: 31

1. Mix the malt with 3½ gallons of water at 165°F or the appropriate temperature to mash at 150°F. Mash for 60 minutes.

2. Recirculate the wort until it's fairly clear. Run off the wort into the kettle.

3. Sparge with 5 more gallons of water at 165°F. Run off the wort into the kettle.

4. Bring the wort to a boil. Boil for 15 minutes. Add the hops and continue to boil for 75 minutes. Add the Whirlfloc tablet at 30 minutes. Put your wort chiller into the wort at least 15 minutes before the end of the boil. Add the fresh chiles during the last 5 minutes of the boil.

5. When the boil finishes, cover the pot with a lid or a new trash bag and chill to 65°F. Siphon the wort into your sanitized fermentor, leaving the chiles behind in the kettle. Pitch two packs of liquid yeast or a 2-liter starter.

6. Ferment at 48°F for 2 days, then raise the temperature to 50°F for 3 more days. Up it to 55°F for another week, then let the temperature rise to 65–68°F for 3 days for a diacetyl rest and to complete fermentation. Crash to 32°F and hold for 7–10 days.

7. Keg or bottle the beer. (If you're bottling, I recommend 4.6 ounces of dextrose/corn sugar for this beer.)

RESOURCES

Homebrew Supply Shops

If there's one near you, support your local homebrew supplier! A local shop not only supplies fresh brewing ingredients, but it's also a good resource for troubleshooting and a great place to meet others in the homebrew community. Homebrew supply shops often are connected closely to local homebrew clubs and give club members discounts on supplies. The American Homebrewers Association (AHA) has a store locator under the Directories tab on its website: **homebrewersassociation.org.**

If you don't have a local shop or you're looking for a specific piece of equipment, try shopping online, where you'll find dozens of stores. If you're buying ingredients, especially hops or yeast, make sure you're ordering from a store with frequent turnover to ensure that everything is fresh. If you're ordering liquid yeast, find a supplier close to you to minimize the time the yeast might spend in a hot delivery truck. Always spend the extra dollar for an optional ice pack.

The suppliers listed here are well respected in the homebrewing community. They're also large enough to have a quick turnover of inventory.

➡ AUSTIN HOMEBREW SUPPLY
austinhomebrew.com
Located in Austin, Texas, Austin Homebrew Supply made its name on the beer kits it produces. Luckily, their shipping prices are very reasonable. Definitely check them out if you're in Texas or nearby states.

➡ MORE BEER!
morebeer.com
While More Beer! has a somewhat limited selection of malt and other ingredients, they make up for it with a huge selection of equipment, much of it fabricated in-house. They're one of the few suppliers who still offer free shipping once you spend a certain amount.

➡ NORTHERN BREWER

northernbrewer.com

One of the largest homebrew suppliers in the country, Northern Brewer has a fine selection of ingredients and supplies. They have retail locations in Minneapolis and St. Paul, Minnesota, and Milwaukee, Wisconsin. It's worth ordering something from them just to get on their mailing list. They put together a really nice catalog and often have promotional codes available only to those on their email list. In the past, they haven't made as much gear as Rebel, More Beer, or Williams, but these days, they're producing more custom products.

➡ REBEL BREWER

rebelbrewer.com

Located in Goodlettsville, Tennessee, Rebel has a huge variety of specialty grains and hard-to-find hops, and they offer bulk discounting on grains. If you live near them, shipping can be pretty affordable, too. Rebel also builds a variety of custom equipment, from temperature controllers to malt mills.

➡ WILLIAMS® BREWING

williamsbrewing.com

Williams is based in San Leandro, California. Check them out if you're in the market for equipment. They have all sorts of distinctive items, from unique wide-brew pots that stretch across two burners for stovetop brewing to nice oxygen diffusion stone wands and 3-gallon kegging kits.

Boutique Yeast Labs

These yeast companies offer dozens of unique strains and blends, and most sell homebrew-size pitches.

➡ EAST COAST YEAST

eastcoastyeast.com

It can prove hard for homebrewers to get their hands on this lab's cultures, though a few homebrew shops do carry them. They're well known for their complex blends of mixed cultures, such as BugFarm and BugCounty.

➡ RVA YEAST LABS

rvayeastlabs.com

They carry a full lineup of ale and lager strains as well as a handful of native strains. I've had good luck with their Piedmont Hops Ale strain, isolated from hop fields in Virginia.

➡ SOUTH YEAST LABS

southyeast.com

This lab has some very interesting cultures sourced from local fruit and flowers. I've had nice results with their N1 strain that they isolated from a Carolina nectarine.

➡ THE YEAST BAY

theyeastbay.com

They specialize in custom Brett blends and mixed cultures as well as some unique ale and lager strains. I'm a big fan of their Amalgamation Brett blend and their Hessian Pils lager yeast. I also really like their Saison Blend 2.

Brewing Calculators and Other Helpful Websites

➡ BEERSMITH™

beersmith.com

The most popular brewing software out there. Most brewers agree that it's well worth the reasonable cost.

➡ BREWER'S FRIEND

brewersfriend.com

This site has one of the most extensive sets of calculators and recipe-formulating programs on the Internet.

➡ EZ WATER CALCULATOR

ezwatercalculator.com

If you want to get geeky with your water's mineral content, check out this site. It's free and very popular among serious brewers.

➡ HOMEBREWING

homebrewing.com/calculators

This site has a good selection of calculators, as well as a nice brewing references section.

➡ THE HOP PAGE

http://realbeer.com/hops

The Hop Page by Glenn Tinseth offers helpful information on growing and using hops. It also features some handy brewing calculators.

➡ THE MAD FERMENTATIONIST

themadfermentationist.com

Michael Tonsmeire stands at the forefront of sour-beer experimentation in homebrewing and in commercial brewing, with Modern Times Beer. His blog is an excellent rabbit hole if you're interested in brewing sours.

➡ MR. MALTY

mrmalty.com

Mr. Malty is the website of Jamil Zainasheff, one of the top homebrewers on the planet, who has won countless awards for his beers. His site is handy for cross-referencing yeast strains from different yeast suppliers and for getting clues as to which commercial brewery originated those strains. It also has a pitch calculator that will tell you how much yeast is the proper amount for any batch.

➡ TASTY BREW

tastybrew.com/calculators

Along with complete recipe information, Tasty Brew has a variety of free programs to help you calculate efficiency, IBUs, and priming sugar weight.

➡ WYEAST LABORATORIES AND WHITE LABS

wyeastlab.com and whitelabs.com

These are two of the largest yeast companies in America, and they both have websites worth exploring. In addition to detailed strain information, their sites feature Q&A sections, tutorials, and more.

Books and Magazines

The number of books available today is mind boggling compared to 25 years ago, when there was only Charlie Papazian's *The Complete Joy of Homebrewing*. Now you can find entire books devoted to particular styles, huge tomes about beers from different countries (Ron Pattinson's *Scotland!* is 758 pages long), and several great magazines.

How to Brew by John Palmer is essential reading for every brewer. It might seem intimidating to beginners, but this book is a go-to reference for the tough questions and chemistry.

With its Brewing Elements Series, Brewers Publications offers a group of reader-friendly books that focus on specific ingredients in beer: *Yeast: The Practical Guide to Beer Fermentation*; *For the Love of Hops: The Practical Guide to Aroma, Bitterness and the Culture of Hops*; and *Water: A Comprehensive Guide for Brewers*. Each book features cutting-edge information about its respective ingredient and is perfect for expanding your brewing knowledge. Brewers Publications also has a multitude of books on individual styles (stout, pale ale, scotch ale, etc.) and genres (wild beer, Belgian beer, etc.). Almost all of them are well worth a read if you're interested in the topic.

Brewing Classic Styles: 80 Winning Recipes Anyone Can Brew by Jamil Zainasheff and John Palmer is the companion book to Zainasheff's podcast on the Brewing Network. The authors cover every style in the Beer Judge Certification Program and go in-depth on the ingredients and techniques needed to brew them.

Designing Great Beers: The Ultimate Guide to Brewing Classic Beer Styles by Ray Daniels has no recipes. Rather, it teaches you how to develop your own. Highly recommended.

If you're interested in brewing sour beers, definitely check out *Wild Brews: Beer Beyond the Influence of Brewer's Yeast* by Jeff Sparrow. He covers all styles of sour beers and offers recipes and techniques for brewing them.

Sacred and Herbal Healing Beers: The Secrets of Ancient Fermentation by Stephen Harrod Buhner covers historical beers that use herbs for medicinal purposes. I'm not sure how drinkable many of the recipes are, but I had fun reading this book.

American Sour Beers by Michael Tonsmeire offers a trove of good information on brewing sour and funky beers.

Brew Your Own: The How-To Homebrew Beer Magazine is a good publication for beginner and intermediate homebrewers.

Zymurgy is the journal of the AHA. It's been around for decades and comes free with a membership to the AHA. Every issue has at least a few articles that are required reading.

Forums

I doubt there's a question about homebrewing that hasn't been answered in an online forum. As with any Internet forum, you have to do a lot of scrolling to get to an informative post, but there's a wealth of information out there.

➡ **THE AMERICAN HOMEBREWERS ASSOCIATION FORUM**
homebrewersassociation.org/forum/index.php
The organization that's been there from the beginning also has its own forum. It's a great resource for homebrewers of any level.

➡ **THE BREWING NETWORK FORUM**
thebrewingnetwork.com/forum
The Brewing Network puts out podcasts (see right) and hosts a forum. Both are great ways to learn and get advice from advanced homebrewers.

➡ **HOMEBREWTALK**
homebrewtalk.com
Plenty of experienced homebrewers share their knowledge and experience in this huge community.

➡ **MILK THE FUNK®**
facebook.com/groups/milkthefunk
This Facebook group with more than 10,000 members has a FAQ section that will answer almost any of your sour or funky beer questions.

Podcasts

Listening to podcasts while you're driving to work or while you're brewing is a great way to learn.

➡ **BASIC BREWING RADIO™**
basicbrewing.com/radio
The programs on Basic Brewing Radio are slightly less entertaining than those on the Brewing Network, but they do stay focused on homebrewing. The shows often feature advanced homebrewers talking about techniques and brewing experiments.

➡ **THE BREWING NETWORK**
thebrewingnetwork.com
The Brewing Network has a variety of programs. Some are rich in information, and others are more focused on entertainment. *The Jamil Show* is the most informative podcast out there; each episode breaks down a particular beer style and tells you how to brew an award-winning example. *Can You Brew It?* is another fun podcast. It features expert homebrewers attempting to clone a commercial beer.

➡ **BREWING TV**
brewingtv.com
This site features an excellent series of instructional videos produced by Northern Brewer (page 190). In addition to brewing how-tos, the videos often offer peeks inside commercial breweries.

➡ **CRAFT BEER RADIO**
craftbeerradio.com
True, Craft Beer Radio is more about beer tasting than brewing, but it's a good listen if you're interested in learning about beer styles and craft brewing in general.

INDEX